# Vortex Simulation and Identification

*Edited by Chaoqun Liu*

Published in London, United Kingdom

Vortex Simulation and Identification
http://dx.doi.org/10.5772/intechopen.104021
Edited by Chaoqun Liu

Assistant to the Editor: Yifei Yu

Contributors
Anthony P. Brown, Chaoqun Liu, Yifei Yu, Alexis Rodriguez Carranza, Obidio Rubio Mercedes, Elder Joel Varas Peréz, Terry E. Moschandreou, Keith C. Afas, Hector Pérez-de-Tejada, Rickard Lundin, Birhanu Abera Kolech



First published in London, United Kingdom, 2024 by IntechOpen
IntechOpen is the global imprint of INTECHOPEN LIMITED, registered in England and Wales, registration number: 11086078, 5 Princes Gate Court, London, SW7 2QJ, United Kingdom

British Library Cataloguing-in-Publication Data
A catalogue record for this book is available from the British Library

Additional hard and PDF copies can be obtained from orders@intechopen.com

Vortex Simulation and Identification
Edited by Chaoqun Liu
p. cm.
Print ISBN 978-1-83768-640-7
Online ISBN 978-1-83768-641-4
eBook (PDF) ISBN 978-1-83768-642-1

For EU product safety concerns:
IN TECH d.o.o., Prolaz Marije Krucifikse Kozulić 3, 51000 Rijeka, Croatia,
info@intechopen.com or visit our website at intechopen.com.

# Meet the editor

Dr. Chaoqun Liu received a BS and MS from Tsinghua University, Beijing, China, in 1968 and 1981, respectively. He obtained a Ph.D. from the University of Colorado at Denver, USA, in 1989. He is currently a Tenured and Distinguished Professor and the director of the Center for Numerical Simulation and Modeling at the University of Texas at Arlington, USA. He has worked on high-order direct numerical simulation/large eddy simulation (DNS/LES) for flow transition and turbulence for 33 years. He is the founder and principal contributor of the Liutex method and third-generation vortex definition and identification methods, including Omega, Liutex/Rortex, Modified Liutex-Omega, Liutex-Core-Line methods, Objective Liutex, RS vorticity decomposition, and UTA R-NR velocity gradient tensor decomposition. He is also the founder of Liutex-based new fluid kinematics.

# Contents

**Preface** **IX**

**Chapter 1** **1**
Periodic Navier Stokes Equations for a 3D Incompressible Fluid with Liutex Vortex Identification Method
*by Terry E. Moschandreou and Keith C. Afas*

**Chapter 2** **23**
Mass Flux in Corkscrew Flow Vortices in the Venus Plasma Wake
*by Hector Pérez-de-Tejada and Rickard Lundin*

**Chapter 3** **33**
Magnetic Skyrmions and Quasi Particles: A Review on Principles and Applications
*by Birhanu Abera Kolech*

**Chapter 4** **61**
Identification of Trailing Vortex Dynamic States
*by Anthony P. Brown*

**Chapter 5** **85**
Numerical Simulation of Energy Cascading in Turbulent Flows Using Sabra Shell Model
*by Alexis Rodriguez Carranza, Obidio Rubio Mercedes and Elder Joel Varas Peréz*

**Chapter 6** **93**
Algorithm to Generate Liutex Core Lines Based on Forward Liutex Magnitude Gradient Lines
*by Yifei Yu and Chaoqun Liu*

# Preface

Vortices are ubiquitous in the universe and include tornados, hurricanes, airplane tip vortices, polar vortices, and even star vortices in the galaxy. Vortices are also building blocks, muscles, and sinews of turbulent flows. A vortex is intuitively recognized as a rotational/swirling motion of fluids, but until recently had no rigidly mathematical definition. In 1858, Helmholtz first defined a vortex composed of so-called vortex filaments, which are infinitesimal vorticity tubes. The vorticity tube is called the first generation of vortex definition and identification, or G1. Although G1 has been accepted by the fluid dynamics community and almost all textbooks for more than a century, we can find many immediate counterexamples. For example, in the laminar boundary layer, where the vorticity (shear) is very large near the wall, but no rotation (no vortex) exists. To solve these contradictions, many vortex criteria methods have been developed during the past four decades. More popular methods are represented by the Q, $\Delta$, $\lambda 2$, and $\lambda ci$ criteria methods. These methods have achieved partial success in vortex identification and are referred to as the second generation of vortex identification, or G2. However, G2 has several critical drawbacks. First, these methods are all scalars that have no rotation axis directions; however, a vortex is a vector. It is hard or impossible to use a scalar to represent a vector. Second, like vorticity, these criteria methods are all contaminated by shear to different degrees. Third, they are all very sensitive to threshold selections. They are also unable to show the vortex structure when both strong and weak vortices coexist. The recently developed Liutex is a third generation of vortex definition and identification, or G3, which is a uniquely defined vector. Liutex has strong potential to be applied to all fluid-related research areas.

Nowadays, the crises human beings face are mainly caused by vortices, like global climate change, polar vortices, tornados, hurricanes, environmental pollution, heart disease, and so on. Therefore, accurate definition and identification of vortices is one of the most challenging research topics for humanity.

The purpose of this book is to encourage all experts who are doing vortex-related research around the world to pay more attention to the progress in recent vortex research.

The book has six chapters covering new vortex theories, vortex identification methods, and vortex simulation and applications.

The editor would like to thank his wife Weilan Jin, his daughter Haiyan Liu, and his son Haifeng Liu for their understanding and unconditional support.

**Chaoqun Liu**
Department of Mathematics,
University of Texas at Arlington,
Arlington, Texas, USA

Chapter 1

# Periodic Navier Stokes Equations for a 3D Incompressible Fluid with Liutex Vortex Identification Method

*Terry E. Moschandreou and Keith C. Afas*


## Abstract

The Incompressible Navier-Stokes Equations (NSEs) are on the list of Millennium Problems, to prove their existence and uniqueness of solutions. The NSEs can be formulated in a periodic 3D domain, where they are termed the Periodic Navier Stokes (PNS) Equations, and can be treated on a subspace spanning a 3-dimensional torus, or $\mathbb{T}^3$. Treating the PNS Equations in $\mathbb{T}^3$-space, this article demonstrates that a decaying of turbulence occurs in the 3D case for the $z$ component of velocity when non-smooth initial conditions are considered for $x, y$ components of velocity and that 'vorticity' sheets in the small scales of 3D turbulence dominate the flow to the extent that non-smooth temporal solutions exist for the $z$ velocity for smooth initial data for the $x, y$ components of velocity. Unlike the Navier-Stokes equations, which have no anti-symmetric vorticity tensor, there are new governing equations which have vorticity tensor and can be decomposed into a rotational part (Liutex), antisymmetric shear and compression and stretching. It is shown that under these recent findings, that there is a strong correlation between vorticity and vorticies for (PNS).


**Keywords:** periodic, Navier-stokes, blow-up, turbulence, 3-torus, Weierstrass, elliptic, analysis

## 1. Introduction

This chapter gives a general model using specific periodic special functions, that is elliptic Weierstrass $P$ functions. The definition of vorticity in [1], is that vorticity is a rotational part added to the sum of antisymmetric shear and compression and stretching. Satisfying a divergence free vector field and periodic boundary conditions respectively with a general spatio-temporal forcing term $f\left(\vec{x}, t\right)$ which is smooth and periodic, then the existence of solutions which blowup in finite time can occur. On the other hand if $u_0$ is not smooth, then there exist globally in time solutions on $t \in [0, \infty)$

with a possible blowup at $t = \infty$. The control of turbulence is possible to maintain when the initial conditions and boundary conditions are posed properly for (PNS) [2–4]. This leads to the following two questions for (PNS),

1. *Is there a decaying of turbulence in the 3D case for the z component of velocity when non-smooth initial conditions are considered for x, y components of velocity?*

   and

2. *Are the vorticity sheets in the small scales of 3D turbulence dominating the flow to the extent that non-smooth temporal solutions exist for the z velocity for smooth initial data for the x, y components of velocity?*

A positive answer exists for both of the above questions [4, 5]. In this chapter it is shown explicitly that for smooth forcing that is both spatial and temporal and Weierstrass P product functions in space for velocity $u_x$ and $u_y$ that the equivalent form of the Navier-Stokes equations derived in [6–8], has as one of the possible solutions for $u_z$ a separable product of spatial functions in the three space variables together multiplied by a general function of $t$ which is a blowup at infinity. On the other hand if $\mathbf{f}\left(\vec{x},t\right)$ is a smooth reciprocal function of a general Weierstrass $P$ function defined on the 3-Torus, then when $u_x$ and $u_y$ in 3D Navier Stokes equations are both in the smooth reciprocal form of the Weierstrass $P$ function then this implies that $u_z$ is not smooth in time. In [6–8] the $z$ component of vorticity was chosen to be constant. Extending the vorticity definition, in particular in this chapter, $u_x, u_y$ satisfy a non-constant spatial or time dependent vorticity for 3D vorticity $\vec{\omega}$. Finally new eqs. [1] are conjectured to possess smooth solutions appearing to not have finite time singularities using the correct definition of vorticity in this study. For (PNS) it is shown that there exists a vortex in each cell of the lattice associated with $\mathbb{T}^3$ using the decomposition of pure rotation(Liutex), antisymmetric shear and compression and stretching. Furthermore it is observed that a singular cusp bifurcation occurs along a principle main axis for the case of smooth and non-smooth initial inputs of velocity.

## 2. Mathematics preliminaries

Let $s \in \mathbb{R}$, the homogenous Sobolev space is,

$$\dot{H}^s\left(\mathbb{T}^3\right) := \left\{ f = \sum_{\mathbf{k} \in \mathbb{Z}^3} a_k e^{i\mathbf{k}\cdot\mathbf{x}}; a_0 = 0 \quad \text{and} \quad \sum_{k \neq 0} |k|^{2s} |a_k|^2 < \infty \right\} \tag{1}$$

with associated norm,

$$\|f\|_{\dot{H}^s} := \left( \sum_{k \neq 0} |k|^{2s} |a_k|^2 \right)^{1/2}$$

The inhomogeneous Sobolev Space is,

$$H^s(\mathbb{T}^3) := \left\{ f = \sum_{\mathbf{k}\in\mathbb{Z}^3} a_k e^{i\mathbf{k}\cdot\mathbf{x}};\ a_0 \neq 0 \quad \text{and} \quad \sum_k |k|^{2s} |a_k|^2 < \infty \right\} \tag{2}$$

with associated norm,

$$\|f\|_{H^s} := \left( \sum_{k\in\mathbb{Z}} |k|^{2s} |a_k|^2 \right)^{1/2} \tag{3}$$

The particular inhomogeneous Sobolev space $H^{\frac{1}{2}}(\mathbb{T}^3)$ is a scale invariant space for (PNS).

Theorem 1 (Preköpa-Leindler) Let $0 < \lambda < 1$ and let $f$, $g$, and $h$ be nonnegative integrable functions on $\mathbb{R}^n$ satisfying,

$$h((1-\lambda)x + \lambda y) \geq f(x)^{1-\lambda} g(y)^{\lambda},$$

for all $x, y \in \mathbb{R}^n$. Then

$$\int_{\mathbb{R}^n} h(x)dx \geq \left( \int_{\mathbb{R}^n} f(x)dx \right)^{1-\lambda} \left( \int_{\mathbb{R}^n} g(x)dx \right)^{\lambda}$$

Theorem 2 (Gagliardo-Nirenberg) Let $1 \leq q \leq \infty$ and $j, k \in \mathbb{N},\ j < k,$ and either,

$$\begin{cases} r = 1 \\ \frac{j}{k} \leq \theta \leq 1 \end{cases} \qquad \begin{cases} 1 < r < \infty \\ k - j - \frac{n}{r} = 0,1,2,\ \dots \\ \frac{j}{k} \leq \theta < 1 \end{cases}$$

If we set $\frac{1}{p} = \frac{j}{n} + \theta\left(\frac{1}{r} - \frac{k}{n}\right) + \frac{1-\theta}{q}$, then there exists constant $C$ independent of $u$ such that

$$\left\| \nabla^j u \right\|_p \leq C \left\| \nabla^k u \right\|_r^{\theta} \|u\|_q^{1-\theta}, \qquad \text{for all} u \in L^q(\mathbb{R}^n) \cap W^{k,r}(\mathbb{R}^n)$$

## 3. Equivalent form of 3D incompressible Navier stokes equations

The 3D incompressible unsteady Navier-Stokes Equations (NSEs) in Cartesian coordinates may be expressed [6–8] as the coupled system Eqs. (4)–(9) below, for the velocity field $\mathbf{u}^* = u^{*i}\vec{\mathbf{e}}_i,\quad u^{*i} = \left\{ u_x^*, u_y^*, u_z^* \right\}$ from the original NSE's,

$$\mathcal{G} = \mathcal{G}_{\delta 1} + \mathcal{G}_{\delta 2} + 3\mathcal{G}_{\delta 3} + \mathcal{G}_{\delta 4} = 0 \tag{4}$$

where,

$$u_x^* \to \frac{u_x}{\delta}, u_y^* \to \frac{u_y}{\delta},\ u_z^* \to \frac{u_z}{\delta}$$
$$x^* \to \delta x,\ y^* \to \delta y,\ z^* \to \delta z,\ t^* \to \delta^2 t$$
$$\frac{\partial}{\partial x^*} \to \delta^{-1}\frac{\partial}{\partial x},\ \frac{\partial}{\partial y^*} \to \delta^{-1}\frac{\partial}{\partial y},\ \frac{\partial}{\partial z^*} \to \delta^{-1}\frac{\partial}{\partial z},\ \frac{\partial}{\partial t^*} \to \delta^{-2}\frac{\partial}{\partial t}$$

and where $\delta = A + 1$ with $A$ being arbitrarily small when $\delta \approx 1$. In [6] $\mathcal{G}$ there is defined without the tensor product term $\mathbf{Q}$. Calculating the tensor product term $\mathbf{Q}$ in Eq. (14) see [6] and using Eq. (22) in [6] shows it's volume integral to approach zero due to $\|\nabla u_z\|_2^2$ approaching zero. See Theorem 1 and 2, where the Preköpa-Leindler and Gagliardo-Nirenberg inequality is used to show this when $r = 1$ and $\lambda = \theta = \frac{1}{2}$. In this paper it is shown that the problem in [6] can be extended to all three velocity components $u_i$. The G.A. decomposition used there shows that there is a missing term (intended to be present) when multiplying Eq. 5 (there) by $u_z$ and adding to the product of $\vec{b}$ and the $z$ momentum equation. This sum is precisely $\|\nabla u_z\|^2$ which is bounded by $\|\nabla u_z\|_2^2$ which is shown below to approach zero as the volume $\Omega$ approaches infinity. Also the pressure due to conservation of forces theorem is a regular function in $t$. As a result it is assumed that it can be written as $P = \tilde{P}(x,y,z)P_{t_0}(t - t_0)$ where $P_{t_0}(t - t_0)$ approaches zero as $t \to t_0$ and $\nabla \cdot \left(\frac{1}{\delta\rho}u_z^2\nabla_{xy}P + \vec{b}\,\frac{1}{\rho}u_z\frac{\partial P}{\partial z}\right) = 3\Phi(t)$ where $\vec{r} = x\vec{i} + y\vec{j} + z\vec{k}$. In Eq. (7) of [6], for the vector $\vec{B} = u_z\nabla \cdot \left(u_z\vec{b}\right)\vec{b}, L\vec{u} = \vec{B}$. Furthermore in Eq. (11) [6] there, the term $\Omega_4$ is the divergence of the vector $\vec{B}$. Using Ostrogradsky's formula in terms of the vorticity $\vec{\omega}$ and velocity $\vec{b}$, $\int_\Omega u_z\vec{b}\nabla \cdot \left(\vec{u}_z\vec{\omega}|\vec{r}|\right)\, d\vec{x} = -\int_\Omega |\vec{r}|u_z\vec{\omega} \cdot \nabla\left(\vec{b}u_z\right)\, d\vec{x}$. Now for a specific pressure $P$ on an $R$-sphere, $\int_\Omega \mathcal{G}_{\delta 1} + \mathcal{G}_{\delta 2} + \mathcal{G}_{\delta 4}\, d\vec{x} = 3\Phi(t)$, where $\Phi(t)$ assumed to be bounded and contain the pressure terms in Eq. (7). The sphere is $|\vec{r}| = R$. Since the 3-Torus is compact there are $m$ closed sets covering it. The outer measure is used where the infimum is taken over all finite subcollections $\mathcal{M}$ of closed spheres $\{E_j\}_{j=1}^n$ covering a specific space associated with $\mathbb{T}^3$. This space is $S_{\mathbb{T}^3}$ and is within $\epsilon$ measure of the 3-Torus and is obtained by minimally smoothening the vertices of $[-L, L]^3$ and slightly puffing out it's facets. Also inner measure is used where the supremum is taken over all finite subcollections $\mathcal{N}$ of closed spheres $\{F_j\}_{j=1}^p$ inside $S_{\mathbb{T}^3}$. Generally by Hölder's inequality, since $\mathcal{G}_{\delta 3}$ is positive for sufficiently sharp increases in radial pressure where $u_z$ does not blowup in finite time (but will be shown to be non-smooth), and for sufficiently small $c > 0$,

$$\sup_{\mathcal{N}}\left|c\left[\int_{F=\bigcup_{j=1}^{p}F_j}\vec{r} \times \nabla_D^{-1}[\mathcal{G}_{\delta 1} + \mathcal{G}_{\delta 2} + \mathcal{G}_{\delta 4}]d\vec{x}\right]\right|$$

$$\leq \inf_{\mathcal{M}}\left|c\int_{\Omega_R=\bigcup_{j=1}^{m}E_j}\vec{r} \times \nabla_D^{-1}[\mathcal{G}_{\delta 1} + \mathcal{G}_{\delta 2} + \mathcal{G}_{\delta 4}]d\vec{x}\right|$$

$$= \left|c\int_\Omega |\vec{r}|\|\nabla_D^{-1}[\mathcal{G}_{\delta 1} + \mathcal{G}_{\delta 2} + \mathcal{G}_{\delta 4}]|\sin(\theta)\vec{n}\ d\vec{x}\right|$$

$$\leq \left|c\int_\Omega |\vec{r}|\|\nabla_D^{-1}[\mathcal{G}_{\delta 1} + \mathcal{G}_{\delta 2} + \mathcal{G}_{\delta 3} + \mathcal{G}_{\delta 4}]|d\vec{x}\right|$$

$$= \left|-c\int_\Omega \left|\vec{r}\right|^2\vec{\omega}_1 \cdot \nabla\left(u_z\vec{b}\right)d\vec{x}\right|$$

$$= c\left|\int_\Omega \left|\vec{r}\right|^2 \vec{\omega}_1 \cdot \nabla\left(u_z\vec{b}\right) d\vec{x}\right|$$

$$\leq c\int_\Omega \left|\left|\vec{r}\right|^2 \vec{\omega}_1 \cdot \nabla\left(u_z\vec{b}\right)\right| d\vec{x}$$

$$\leq c\left(\int_\Omega \left|\vec{r}\right|^4 d\vec{x}\right)^{\frac{1}{4}}\left[\left(\int_\Omega \left|\vec{w}_1 \cdot \nabla\left(u_z\vec{b}\right)\right|^2\right)^{\frac{3}{4}}\right]$$

$$\leq c\left(\int_\Omega \left|\vec{r}\right|^4 d\vec{x}\right)^{\frac{1}{4}}\left[\left[\left(\int_\Omega \left|\vec{\omega}_1 \cdot \left(\nabla u_z \otimes \vec{b}\right)\right|^2 d\vec{x}\right)^{1/2}\right]^{\frac{3}{2}}\right.$$

$$\left.+\left[\left(\int_\Omega \left|u_z\nabla\left(\vec{b}\right) \cdot \vec{\omega}_1\right|^2 d\vec{x}\right)^{1/2}\right]^{\frac{3}{2}}\right]$$

$$\leq c\left(\int_\Omega \left|\vec{r}\right|^4 d\vec{x}\right)^{\frac{1}{4}}\left[\left[\int_\Omega \left|\vec{\omega}_1\right|^2\left|\vec{u}\right|^2\left|\nabla u_z\right|^2 d\vec{x}\right]^{\frac{3}{4}}\right.$$

$$\left.+\left[\int_\Omega \left|u_z\right|^2\left|\nabla\vec{b}\right|^2\left|\vec{\omega}_1\right|^2 d\vec{x}\right]^{\frac{3}{4}}\right]$$

$$\leq c\left(\int_\Omega \left|\vec{r}\right|^4 d\vec{x}\right)^{\frac{1}{4}}\left[\left[\left\|\vec{\omega}_1\right\|_4^2\left\|\vec{u}\right\|_8^2\left\|\nabla u_z\right\|_{16}^2\right]^{\frac{3}{4}}\right.$$

$$\left.+\left[\int_\Omega \left|u_z\right|^2\left|\nabla\vec{b}\right|^2\left|\vec{\omega}_1\right|^2 d\vec{x}\right]^{\frac{3}{4}}\right]$$

$$\leq \frac{152}{15}cL^{7/4}\left[\frac{1}{c}\left\|\vec{\omega}_1\right\|_4^{\frac{3}{2}}\left\|\vec{u}\right\|_8^{\frac{3}{2}}\left\|\nabla u_z\right\|_2^3 + \left[\frac{1}{c}\int_\Omega \left\|u_z\right\|_8^{\frac{3}{2}}\left\|\nabla\vec{b}\right\|_2^3\left\|\vec{\omega}_1\right\|_4^{\frac{3}{2}} d\vec{x}\right]^{\frac{3}{4}}\right]$$

$$\boxed{\leq 0(\text{due to } 2-\text{norms approaching zero}}$$

Considering inscribed spheres in each puffed cell of the perturbed lattice $S_{\mathbb{T}^3}$ and integrating over the union of all such spheres where each cell is of arbitrary small measure containing spheres results in a finite well defined integral bounded by a now finite supremum. The supremum on the left most side of above chain of inequalities may exist for certain Elliptic functions defined on the 3-Torus with limiting parameters as the present study aims to show. Then integration over $\mathbb{T}^3$ is well defined and we can replace $S_{\mathbb{T}^3}$ by $\mathbb{T}^3$. Now $\vec{\omega}_1$ is vorticity $\vec{\omega}$ multiplied with $u_z$ and to begin with $\nabla_D^{-1}$ is the inverse of the divergence operator integrated over an arbitrary cell of the 3-Torus lattice. Here the identity $A \times B = |A||B|\sin(\theta)\vec{n}$ was used and in the previous two integrals over a periodic lattice, the first one approaches zero as $R$ becomes arbitrarily large, where the 3-Torus being a compact manifold is bounded by an $R$-Ball with the norm of the gradient of $u_z$ approaching zero and the second integral consisting of $|\nabla\vec{b}|$ approaches zero for the finite but large volume 3-Torus. The Preköpa-Leindler and Gagliardo-Nirenberg Inequalities have been used for each

$$|\Omega|^{\frac{1}{2}}\|\nabla u_j\|_2 \leq C\|u_j\|_q^{1-\theta}\left[|\Omega|^{\frac{1}{2}}\|\nabla^2 u_j\|_r^{\theta}\right] \leq C\|u_j\|_q^{1-\theta}\int_{\Omega}\nabla^2 u_j d\vec{x}$$

Here $\sum_{j=1}^{3} u_j = \Psi * w$, where * is convolution and $\int_{\mathbb{T}^3} w \; d\mathbf{x} = \mathbf{s} \in \mathbb{R}$. If $\Psi$ is the fundamental solution of the scalar Laplacian on the 3-Torus $\mathbb{T}^3 = S^1 \times S^1 \times S^1$ noting that $\Delta(\Psi * w) = w$ then the integral of the Laplacian is in general non zero [9]. We must rely on the dimension of the Lattice to ensure the limit value is zero upon dividing by large enough $|\Omega|$. Off of the associated compact set the velocity is zero or the velocity has compact support. The chain of inequalities at top of this page imply that, in general $\nabla_D^{-1}(\mathcal{G}_{\delta 1} + \mathcal{G}_{\delta 2} + \mathcal{G}_{\delta 4}) = \vec{r}\Phi(t)$ since the two vectors $\vec{r}$ and $\nabla_D^{-1}$ can be in the same direction. A term $\Phi(t)$ is also multiplied by $\vec{r}$. A group $G$ of transformations of $\vec{u}\left(\vec{r},t\right)$ is a symmetry group of NS if for all $g \in G$, $\vec{u}$ a NS solution implies $g\vec{u}$ is a NS solution. The group is $\mathbb{Z}$. The Navier Stokes equations are invariant under the dilation group as shown after Eq. (4). Next there will be an application of the group transformations seen in Eqs. (5)–(8):

$$\mathcal{G}_{\delta 1} = -\left(1-\frac{1}{\delta}\right)\left(\frac{\partial u_z}{\partial t}\right)\left(\frac{\partial u_z}{\partial t} - \frac{\mu}{\rho}\nabla^2 u_z + \frac{1}{\rho}\frac{\partial P}{\partial z}\right) \tag{5}$$

$$\mathcal{G}_{\delta 2} = u_z\frac{\partial u_z}{\partial z}\frac{\partial u_z}{\partial t} + u_z^2\frac{\partial^2 u_z}{\partial z \partial t} + \frac{2u_z}{\delta}\left(\frac{\partial \mathbf{u}}{\partial t}\cdot\nabla u_z\right) \tag{6}$$

$$\mathcal{G}_{\delta 3} = \iint_{\partial\Omega}\frac{u_z}{\rho}\left(\frac{1}{\delta}u_z\nabla_{xy}P + \frac{\partial P}{\partial z}\vec{b}\right)\cdot\vec{n}\; dS - \int_{\Omega}\left\|\left(\frac{\vec{b}}{\|\vec{b}\|}\cdot\nabla u_z\right)\frac{\partial u_z}{\partial t}\vec{b}\right\| dV \tag{7}$$

$$\mathcal{G}_{\delta 4} = \delta^2\vec{F}_T\cdot\nabla u_z^2 - \delta^3 u_z\frac{\partial u_z}{\partial z}F_z + \delta^3\vec{b}\cdot\nabla(u_z F_z) + \delta^3\left(1-\frac{1}{\delta}\right)\frac{F_z}{u_z}\frac{\partial u_z}{\partial t} \tag{8}$$

where $\vec{b} = \frac{1}{\delta}\left(u_x\vec{i} + u_y\vec{j} + u_z\vec{k}\right)$, and $\vec{i}, \vec{j}$ and $\vec{k}$ are the standard unit vectors. For Poisson's Equation seen in Eq. (9) (see [6, 8]), the second derivative $P_{zz}$ is set equal to the second derivative obtained in the $\mathcal{G}_{\delta 1}$ expression as part of $\mathcal{G}$,

$$\begin{aligned} P_{zz} = -2u_z\nabla^2 u_z - \left(\frac{\partial u_z}{\partial z}\right)^2 + \frac{1}{\eta}\frac{\partial}{\partial z}\left(\frac{\partial u_z}{\partial x} + \frac{\partial u_z}{\partial y}\right) - \delta u_x\frac{\partial^2 u_z}{\partial z \partial x} - \\ \delta u_y\frac{\partial^2 u_z}{\partial z \partial y} + \left(\frac{\partial u_x}{\partial x}\right)^2 + 2\frac{\partial u_x}{\partial y}\frac{\partial u_y}{\partial x} + \left(\frac{\partial u_y}{\partial y}\right)^2 \end{aligned} \tag{9}$$

where the last three terms on rhs of Eq. (9) can be shown to be equal to $-(P_{xx} + P_{yy})$. Along with Eqs. (4)–(9), the continuity equation in Cartesian coordinates, is $\nabla^i u_i = 0$. Furthermore the right hand side of the one parameter group of transformations are next mapped to $\eta$ variable terms,

$$u_i = \frac{1}{\eta}v_i, \quad P = \frac{1}{\eta^2}Q, \quad x_i = \eta y_i, \quad t = \eta^2 s, \qquad i = 1,2,3. \tag{10}$$

and Eq. (4) becomes,

$$\mathcal{G}(\eta) = \mathcal{G}(\eta)_{\delta 1} + \mathcal{G}(\eta)_{\delta 2} + \mathcal{G}(\eta)_{\delta 3} + \mathcal{G}(\eta)_{\delta 4} = 0 \tag{11}$$

where,

$$\mathcal{G}(\eta)_{\delta1} = \frac{1}{\eta^6}\left[(\delta^{-1}-1)\left(\frac{\partial v_3}{\partial s}\right)^2 + \frac{\mu\left(\frac{\partial v_3}{\partial s}\right)\left(\frac{\partial^2 v_3}{\partial {y_1}^2}+\frac{\partial^2 v_3}{\partial {y_2}^2}+\frac{\partial^2 v_3}{\partial {y_3}^2}\right)}{\rho}(1-\delta^{-1}) + \frac{(\delta^{-1}-1)\left(\frac{\partial v_3}{\partial s}\right)\frac{\partial Q}{\partial y_3}}{\rho}\right] \tag{12}$$

$$\mathcal{G}(\eta)_{\delta2} = \frac{v_3}{\eta^6}\left(\frac{\partial v_3}{\partial y_3}\right)\frac{\partial v_3}{\partial s} + \frac{(v_3)^2}{\eta^6}\frac{\partial^2 v_3}{\partial y_3 \partial s} + \frac{2\left(\frac{\partial v_1}{\partial s}\right)v_3\frac{\partial v_3}{\partial y_1} + 2\left(\frac{\partial v_2}{\partial s}\right)v_3\frac{\partial v_3}{\partial y_2} + 2\left(\frac{\partial v_3}{\partial s}\right)v_3\frac{\partial v_3}{\partial y_3}}{\delta\eta^6} \tag{13}$$

$$\mathcal{G}(\eta)_{\delta3} = \frac{1}{\eta^3}\times\left[\iint_S\left(\frac{1}{\delta\rho}v_3^2\nabla_{y_1y_2}Q + \frac{1}{\delta}\vec{v}\frac{1}{\rho}v_3\frac{\partial Q}{\partial y_3}\right)\cdot\vec{n}\ dS - \int_\Omega \frac{\|\frac{\partial v_3}{\partial s}\vec{b}\cdot\left(\vec{b}\otimes\nabla v_3\right)\|}{\|\vec{b}\|}dV\right] \tag{14}$$

$$\mathcal{G}(\eta)_{\delta4} = \frac{1}{\eta^3}\left[\delta^2\vec{F}_T\cdot\nabla_{y_1y_2}v_3^2 - \delta^3 v_3\frac{\partial v_3}{\partial y_3}F_z + \delta^2\vec{v}\cdot\nabla(v_3F_z)\right] \tag{15}$$

where $\vec{v} = (v_1, v_2, v_3)$ and $\vec{F}_T = F_{T_1}\vec{i} + F_{T_2}\vec{j}$. The body force $\mathbf{F}^* = F^{*i}\mathbf{e}_i$, with $\vec{F}_T = (F_{T_1}(y_1,y_2,y_3,s), F_{T_2}(y_1,y_2,y_3,s))$ and $F_z$ is the z–component of the force vector. $P$ depends on $\eta$ as $P = \frac{1}{\eta^2}Q$. Thus $\frac{\partial P}{\partial z} = \frac{1}{\eta^3}\frac{\partial Q}{\partial y_3}$. We solve for $\frac{\partial P}{\partial z}$ and using Poisson's equation Eq. (9), set second derivatives of P w.r.t. $z$ equal to each other, and then set $\delta \approx 1$ after multiplying a factor of $\delta - 1$ out of the equation. This makes $A = \delta - 1$, a (small) perturbation parameter. Considering the Kinematic Viscosity $\nu = \mu/\rho$, since it was shown that there exists a $C$ such that $\left(\frac{\partial u_z}{\partial t}\right)^2 = C^2\nu^2\left[\frac{\|\nabla u_z\|_2^2}{u_z}\right]^2$ in earlier work ([6], page 392) then since it was also demonstrated $\|\nabla u_z\|_2^2 = O(\epsilon\eta) = \frac{\eta}{\|\Omega\|}$, due to increasing measure of $\Omega$, it can be seen that $\left(\frac{\partial u_z}{\partial t}\right)^2 = \frac{\nu^2\epsilon^2\eta^2}{\eta^2 v_3^2} = \frac{\nu^2\epsilon^2}{v_3^2} = \frac{\zeta^2}{v_3^2}$. This implies that the constant $C$ is given as in [10]:

$$C = \inf\frac{\int_\Omega \|\nabla u\|^2 dv\left(\int_\Omega u^2 dv\right)^{2/d}}{\int_\Omega \|u - u_\Omega\|^{2+4/d}dv}$$

where $d = 3$ and the infimum is taken over functions $u \in W^{1,1}(\Omega)$ and $u_\Omega$ is its average $\|\Omega\|^{-1}\int_\Omega u dv$. Calculation of above integration terms leads to the identification that $C \approx O(1)$, giving $\eta^6$ order when transforming with Eq. (10) in $C^2\nu^2\left[\frac{\|\nabla u_z\|_2^2}{u_z}\right]^2$. The final operators become independent of $\eta$ and the equation is in the form,

$$\mathcal{L} = \mathcal{L}_1 + \mathcal{L}_2 + \frac{\eta^3}{\epsilon}\mathcal{L}_3 + \frac{\eta^3}{\epsilon}\mathcal{L}_4 = 0 \tag{16}$$

where the four components are defined as:

$$\mathcal{L}_1 = \nu^2\epsilon^2(\delta^{-1}-1)v_3^{-2} + (1-\delta^{-1})\left(\frac{\partial v_3}{\partial s}\right)\left(\frac{\partial^2 v_3}{\partial {y_1}^2}+\frac{\partial^2 v_3}{\partial {y_2}^2}+\frac{\partial^2 v_3}{\partial {y_3}^2}\right) + (\delta^{-1}-1)\frac{\left(\frac{\partial v_3}{\partial s}\right)\frac{\partial Q}{\partial y_3}}{\rho} \tag{17}$$

$$\mathcal{L}_2 = v_3\left(\frac{\partial v_3}{\partial y_3}\right)\frac{\partial v_3}{\partial s} + (v_3)^2\frac{\partial^2 v_3}{\partial y_3 \partial s} + \frac{1}{\delta}\left[2\left(\frac{\partial v_1}{\partial s}\right)v_3\frac{\partial v_3}{\partial y_1} + 2\left(\frac{\partial v_2}{\partial s}\right)v_3\frac{\partial v_3}{\partial y_2} + 2\left(\frac{\partial v_3}{\partial s}\right)v_3\frac{\partial v_3}{\partial y_3}\right] \tag{18}$$

$$\mathcal{L}_3 = \iint_S \left(\frac{1}{\delta\rho}v_3^2\nabla_{y_1y_2}Q + \frac{1}{\delta}\vec{v}\frac{1}{\rho}v_3\frac{\partial Q}{\partial y_3}\right)\cdot\vec{n}\ dS - \int_\Omega \frac{\|\frac{\partial v_3}{\partial s}\vec{b}\cdot\left(\vec{b}\otimes\nabla v_3\right)\|}{\|\vec{b}\|}dV \tag{19}$$

$$\mathcal{L}_4 = \delta^2\vec{F}_T\cdot\nabla_{y_1y_2}v_3^2 - \delta^3 v_3\frac{\partial v_3}{\partial y_3}F_z + \delta^2\vec{v}\cdot\nabla(v_3F_z) \tag{20}$$

It can be seen that the expressions in square brackets in Eqs. (14) and (15) are $\mathcal{L}_3$ & $\mathcal{L}_4$ in Eqs. (19) and (20), respectively. Finally, in subsequent sections the Weierstrass degenerate elliptic P function will be used. Letting the Weierstrass P function be denoted by $\wp(z, g2, g3)$, the degenerate case can be denoted as $P_m(z) = \wp(z, 3m^2, m^3)$.

### 3.1 Decomposition of NSEs

For Eqs. (4)–(9) the Dirichlet condition $\vec{u}^*\left(\vec{x}^*, 0\right) = \vec{\xi}\left(\vec{x}^*\right)$ such that $\nabla\cdot\vec{\xi} = 0$ describes the NSEs together with an incompressible initial condition. Considering periodic boundary conditions defined on 3-torus with associated Lattice is a periodic BVP for the NSEs. Solutions were found to be in the form,

$$\mathbf{u} = \left(u_x, u_y, u_z\right) : \mathbb{R}^+ \times \mathbb{R}^3/\mathbb{Z}^3 \to \mathbb{R}^3 \tag{21}$$

where $u_x, u_y$ and $u_z$ satisfy Eqs. (4)–(9).

### 3.2 Liutex vector and respective governing equations

Theorem 1 in [6] is used in the above decomposition of (PNS) and is the basis theorem of this paper. By using the generalized divergence theorem for scalar products, the term $\nabla u_z^2\cdot\frac{\partial\vec{b}}{\partial t}$ in Eq. (6) is extended to 3-components of velocity. (see Eq. (13) in [8] which only incorporated a 2-component velocity field). No finite time blowup was obtained for the simplified case there. Solutions are obtained symbolically with Maple 2021 software, and with the use of Poisson's equation, Eqs. (4)–(9), lead to,

$$L = L_1 + 2(L_2 + L_3)\frac{\partial v_3}{\partial s} = 0$$

with $L_1, L_2$ and $L_3$ expressions given in Appendix 1. Solving symbolically for $L = 0$ individually for the mixed partial derivatives in the expression $\left(\frac{\partial^2 v_2}{\partial y_3\partial s} - \frac{\partial^2 v_1}{\partial y_3\partial s}\right)$, using the following definition of $\kappa$ in terms of $v_1, v_2$ and $v_3$, [see Eq. (33) in [1]. The new equation there has a viscous term reduced to half and one only needs to calculate 3 elements.]

$$-\left(\frac{\partial^2 v_2}{\partial y_3\partial s} - \frac{\partial^2 v_1}{\partial y_3\partial s}\right) = \frac{\partial}{\partial s}\kappa(y_1, y_2, y_3, s) - \frac{\partial^2 v_3}{\partial y_2\partial s} + \frac{\partial^2 v_3}{\partial y_1\partial s} \tag{22}$$

Both,

$$M_1 = \frac{\partial^2 v_1}{\partial y_3 \partial s} \tag{23}$$

and

$$M_2 = \frac{\partial^2 v_2}{\partial y_3 \partial s} \tag{24}$$

are nonlinear partial differential equations. The $v_1$ and $v_2$ velocities are chosen respectively as the following general spatial–temporal functions, which are assumed to fulfill compatibility conditions in [11, 12],

$$v_1(y_1, y_2, y_3, s) = (U_1(y_1, y_2, y_3) + A) \times f_1(s) \tag{25}$$

and

$$v_2(y_1, y_2, y_3, s) = (U_2(y_1, y_2, y_3) + A) \times f_2(s) \tag{26}$$

where $A \ll 1$ and positive. Note that the magnitude of Liutex (scalar form) is obtained in the plane perpendicular to the local axis, which is twice the angular speed of local fluid rotation,

$$\vec{\omega}_L = \frac{2}{\left|\vec{r}\right|^2}\left(\vec{r} \times \vec{v}\right) \tag{27}$$

where $\omega_L$ is associated with the Liutex vector part of vorticity, $r = y_1\vec{i} + y_2\vec{j} + y_3\vec{k}$ and where the Liutex magnitude difference is calculated as follows,

$$\kappa(y_1, y_2, y_3, s) = 2\frac{y_2 v_3 - y_3 v_2 - y_1 v_3 + y_3 v_1}{y_1{}^2 + y_2{}^2 + y_3{}^2} \tag{28}$$

Substituting $M_1 - M_2$ expressions in Eqs. (23) and (24) and $\kappa$ from Eq. (28) into Eq. (22) gives a new PDE,

$$\mathcal{F} = \frac{\mathcal{F}_1}{H} + 2\frac{\mathcal{F}_2}{H} + \mathcal{F}_3 = 0, \tag{29}$$

$$H = \left(\frac{\partial v_3}{\partial s}\right) v3 \left(\frac{\partial v_3}{\partial y1}\right) \frac{\partial v_3}{\partial y2}$$

$$\begin{aligned}\mathcal{F}_1 = \bigg[&\frac{1}{2}\left(\frac{\partial v_3}{\partial s}\right) v_3 \frac{\partial^3 v_3}{\partial y_3{}^2 \partial s} + \left(\frac{\partial^2 v_3}{\partial y_3 \partial s} v_3\right)\left(-\frac{1}{2} v_3 \frac{\partial^2 v_3}{\partial y_3 \partial s} + (-F_{T_1} - U_1)\frac{\partial v_3}{\partial y_1} v_3 + \right.\\ &\left.(-U_2 - F_{T_2})\frac{\partial v_3}{\partial y_2} - \frac{\Phi(s)}{2} + \left(\frac{\partial v_3}{\partial s}\right)\frac{\partial v_3}{\partial y_3}\right)\bigg] \times \left(\frac{\partial v_3}{\partial y_1} - \frac{\partial v_3}{\partial y_2}\right)\end{aligned} \tag{30}$$

$$\mathcal{F}_2=\left[v_3\left(\frac{\partial v_3}{\partial y_1}-\frac{\partial v_3}{\partial y_2}\right)(F_{T_1}+U_1)\frac{\partial^2 v_3}{\partial y_3\partial y_1}+v_3\left(\frac{\partial v_3}{\partial y_1}-\frac{\partial v_3}{\partial y_2}\right)(F_{T_2}+U_2)\frac{\partial^2 v_3}{\partial y_3\partial y_2}+\right.$$
$$\frac{3}{2}\left(\frac{\partial v_3}{\partial s}\right)v_3\left(\frac{\partial v_3}{\partial y_1}-\frac{\partial v_3}{\partial y_2}\right)\frac{\partial^2 v_3}{\partial {y_3}^2}+\frac{3}{2}\left(\frac{\partial v_3}{\partial y_3}\right)^2\left(\frac{\partial v_3}{\partial y_1}-\frac{\partial v_3}{\partial y_2}\right)\frac{\partial v_3}{\partial s}+$$
$$\left((F_{T_1}+U_1)\frac{\partial v_3}{\partial y_3}+v_3\left(\frac{\partial U_1}{\partial y_3}U_1+\frac{\partial F_{T_1}}{\partial y_3}\right)\right)\left(\frac{\partial v_3}{\partial y_1}\right)^2-\left(\frac{\partial v_3}{\partial y_2}\right)\left((U_1-U_2+F_{T_1}-F_{T_2})\frac{\partial v_3}{\partial y_3}+\right.$$
$$\left.v_3\left(\frac{\partial F_{T_1}}{\partial y_3}-\frac{\partial F_{T_2}}{\partial y_3}\right)\right)\frac{\partial v_3}{\partial y_1}-\left(\frac{\partial v_3}{\partial y_2}\right)^2\left((F_{T_2}+U_2)\frac{\partial v_3}{\partial y_3}+v_3\left(\frac{\partial U_2}{\partial y_3}+\frac{\partial F_{T_2}}{\partial y_3}\right)\right)\Bigg]\frac{\partial v_3}{\partial s} \tag{31}$$

$$\mathcal{F}_3=-\left[\frac{(-2y_1+2y_2)\frac{\partial v_3}{\partial s}+2y_3(U_1(y_1,y_2,y_3)-U_2(y_1,y_2,y_3))}{{y_1}^2+{y_2}^2+{y_3}^2}-\frac{\partial^2 v_3}{\partial y_2\partial s}+\frac{\partial^2 v_3}{\partial y_1\partial s}\right] \tag{32}$$

$$\begin{aligned}F_{T_1}(y_1,y_2,y_3,s)&=f_0(s)(F(y_1,y_2,y_3)+A)\\F_{T_2}(y_1,y_2,y_3,s)&=f_0(s)(G(y_1,y_2,y_3)+A)\end{aligned} \tag{33}$$

### 3.3 Case 1

An example of a smooth force $f_0(s)$ at $s=s_0+\alpha$ is considered with it's accompanying solution of Eq. (29),

$$\begin{aligned}&f_0(s)=\operatorname{sech}(s-s_0)^2\\&\frac{dF_4(s)}{ds}=\sqrt[3]{3c_4\int\Phi(s)\,ds+C_1},\qquad \text{smooth}\quad f_0(s)\end{aligned} \tag{34}$$

For small $m \ll 1$,

$$\begin{aligned}F(y_1,y_2,y_3)&=\mathcal{P}(y_1,3m^2,m^3)^{-1}\times\mathcal{P}(y_2,3m^2,m^3)^{-1}\times\mathcal{P}(y_3,3m^2,m^3)^{-1}\\G(y_1,y_2,y_3)&=\mathcal{P}(y_1,3m^2,m^3)^{-1}\times\mathcal{P}(y_2,3m^2,m^3)^{-1}\times\mathcal{P}(y_3,3m^2,m^3)^{-1}\end{aligned} \tag{35}$$

Setting $v_1$ and $v_2$, as

$$v_1=f_{a_1}(s)\left(\mathcal{P}(y_1,3m^2,m^3)^{-1}\times\mathcal{P}(y_2,3m^2,m^3)^{-1}\times\mathcal{P}(y_3,3m^2,m^3)^{-1}+A\right)$$

$$v_2=f_{a_2}(s)\left(\mathcal{P}(y_1,3m^2,m^3)^{-1}\times\mathcal{P}(y_2,3m^2,m^3)^{-1}\times\mathcal{P}(y_3,3m^2,m^3)^{-1}+A\right) \tag{36}$$

where as an example $f_{a_1}(s)=f_{a_2}(s)=\tanh(s-s_0)$, a tan hyperbolic linearization in $s$ at $s_0$. Here the relationship between $f_0$ and each of $v_i$ is $f_0=\frac{d}{ds}f_{a_i}.\mathcal{P}^{-1}$ is the reciprocal of the degenerate Weierstrass P function with parameter $m$ plotted for some $m$ values listed in captions in **Figure 1**. The definition of degenerate function is,

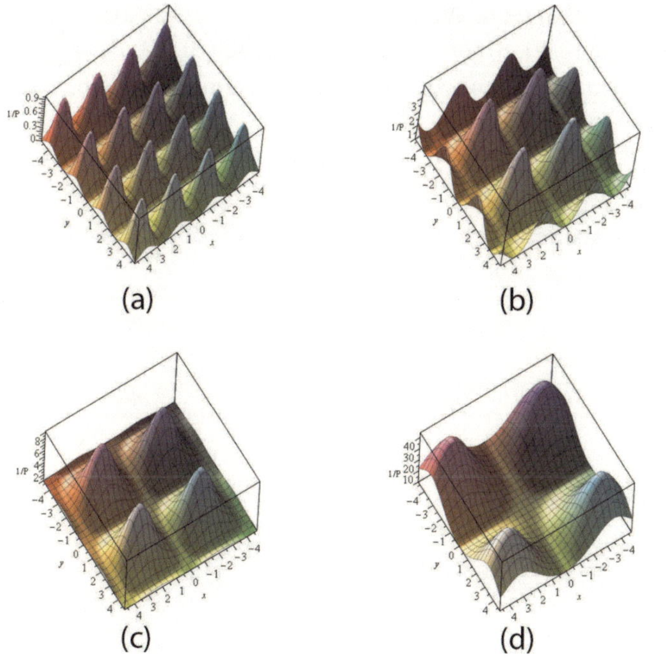


**Figure 1.**
*Plots of the reciprocal of the degenerate Weierstrass P functions in two dimensions $(y_1, y_2)$ given relative to the canonical Weierstrass P functions $\wp(y_i, g2, g3)$ as $\mathcal{P}_{m=n}(y_i) = \wp(y_i, 3n^2, n^3)$. (a) Reciprocal of Weierstrass degenerate P function for $g2 = 3m^2$, $g3 = m^3$, $m = 1$. (b) Reciprocal of Weierstrass degenerate P function for $g2 = 3m^2$, $g3 = m^3$, $m = \frac{1}{2}$. (c) Reciprocal of Weierstrass degenerate P function for $g2 = 3m^2$, $g3 = m^3$, $m = \frac{7}{20}$. (d) Reciprocal of Weierstrass degenerate P function for $g2 = 3m^2$, $g3 = m^3$, $m = \frac{3}{20}$.*

$$\mathcal{P}(z, 3m^2, m^3) = -\frac{m}{2} + \frac{3}{2} m \csc\left(\frac{z\sqrt{6}\sqrt{m}}{2}\right)^2 \tag{37}$$

### 3.4 Case 2

An example of a smooth force $f_0(s)$ at $s = s_0$ is considered with it's accompanying solution of Eq. (29),

$$\begin{aligned} f_0(s) &= \operatorname{sech}(s - s_0)^2 \\ \frac{dF_4(s)}{ds} &= c_4 \frac{df_0(s)}{ds} \qquad \text{smooth,} \quad f_0(s) \end{aligned} \tag{38}$$

Setting $v_1$ and $v_2$, as

$$\begin{aligned} v_1 &= f_{a_3}(s)\mathcal{P}(y_1, 3m^2, m^3) \times \mathcal{P}(y_2, 3m^2, m^3) \times \mathcal{P}(y_3, 3m^2, m^3) \\ v_2 &= f_{a_4}(s)\mathcal{P}(y_1, 3m^2, m^3) \times \mathcal{P}(y_2, 3m^2, m^3) \times \mathcal{P}(y_3, 3m^2, m^3) \end{aligned} \tag{39}$$

$$f_{a_3}(s) = -f_{a_4}(s) \tag{40}$$

where $\mathcal{P}$ is the degenerate Weierstrass P function with parameter $m$. Also for case 2, $\Phi(s) = 0$ and the relationship between $f_0$ and each of $v_i$ is $f_0 = \frac{d}{ds} f_{a_i}$. The general reducing solution in Section 4 will also be considered for the two cases $\Phi(s) = 0$ and $\Phi(s) \neq 0$. It follows that $\Phi(s) = 0$ iff $v_1 = -v_2$, from which it follows that $\iint_S v_3^2 \left( n_3 \frac{\partial P}{\partial y_3} + \left( \nabla_{y_1 y_2} P \cdot \vec{n} \right) \right) dS = 0$ and the slope of the time dependent linear solution of $v_3$ is arbitrarily small.

For case 2, for spatially non-smooth $v_1$ and $v_2$, the solution of Eq. (29) is in the form $v_3 = F_4(s) F_5(y_1, y_2, y_3)$ which satisfies,

$$\frac{\mathrm{d}}{\mathrm{d}t} F_4(s) = c_4 \frac{df_0}{ds} \tag{41}$$

$$\begin{aligned}
&(P(y_2, 3m^2, m^3))^2 \left( \left( \frac{\partial^2 F_5}{\partial y_3 \partial y_1} \right) F_5 + \left( \frac{\partial^2 F_5}{\partial y_3 \partial y_2} F_5 \right) F_5 + 1/2 \left( \frac{\partial F_5}{\partial y_3} \right) \left( \frac{\partial F_5}{\partial y_1} + \frac{\partial F_5}{\partial y_2} \right) \right) \times \\
&(P(y_1, 3m^2, m^3))^2 ({y_1}^2 + {y_2}^2 + {y_3}^2) \left( \frac{\partial F_5}{\partial y_1} - \frac{\partial F_5}{\partial y_2} \right) (P(y_3, 3m^2, m^3))^3 + P(y_2, 3m^2, m^3) \\
&\times \Bigg( P(y_2, 3m^2, m^3) P'(y_3, 3m^2, m^3) \left( \frac{\partial F_5}{\partial y_1} - \frac{\partial F_5}{\partial y_2} \right) \left( \frac{\partial F_5}{\partial y_1} + \frac{\partial F_5}{\partial y_2} \right) \\
&\times ({y_1}^2 + {y_2}^2 + {y_3}^2) P(y_1, 3m^2, m^3) + 1/4 \Bigg( 7 F_5 ({y_1}^2 + {y_2}^2 + {y_3}^2) \left( \frac{\partial F_5}{\partial y_1} - \frac{\partial F_5}{\partial y_2} \right) \frac{\partial^2 F_5}{\partial {y_3}^2} \\
&-2 \left( \frac{\partial F_5}{\partial y_2} \right) ({y_1}^2 + {y_2}^2 + {y_3}^2) \left( \frac{\partial F_5}{\partial y_1} \right)^2 + 7 \Bigg( (2/7 {y_1}^2 + 2/7 {y_2}^2 + 2/7 {y_3}^2) \left( \frac{\partial F_5}{\partial y_2} \right)^2 \\
&+4/7 F_5 (y_1 - y_2) \frac{\partial F_5}{\partial y_2} + \left( \frac{\partial F_5}{\partial y_3} \right)^2 ({y_1}^2 + {y_2}^2 + {y_3}^2) \Bigg) \frac{\partial F_5}{\partial y_1} \\
&-7 \left( \frac{\partial F_5}{\partial y_2} \right) \left( \frac{\partial F_5}{\partial y_3} \right)^2 ({y_1}^2 + {y_2}^2 + {y_3}^2) \Bigg) c_4 \Bigg) F_5 \times P(y_1, 3m^2, m^3) (P(y_3, 3m^2, m^3))^2 \\
&- \left( \left( \frac{\partial^2 F_5}{\partial y_3 \partial y_1} \right) F_5 + \left( \frac{\partial^2 F_5}{\partial y_3 \partial y_2} \right) F_5 + 1/2 \left( \frac{\partial F_5}{\partial y_3} \right) \left( \frac{\partial F_5}{\partial y_1} + \frac{\partial F_5}{\partial y_2} \right) \right) ({y_1}^2 + {y_2}^2 + {y_3}^2) \\
&\times \left( \frac{\partial F_5}{\partial y_1} - \frac{\partial F_5}{\partial y_2} \right) P(y_3, 3m^2, m^3) + P'(y_3, 3m^2, m^3) F_5 \left( \frac{\partial F_5}{\partial y_1} - \frac{\partial F_5}{\partial y_2} \right) \left( \frac{\partial F_5}{\partial y_1} + \frac{\partial F_5}{\partial y_2} \right) \\
&\times ({y_1}^2 + {y_2}^2 + {y_3}^2) = 0
\end{aligned} \tag{42}$$

The solution of Eq. (41) for $F_4(s)$ is,

$$F_4 = c_4 f_0(s) + C_1 \tag{43}$$

where it is observed that it is smooth at $s = s_0$ and the limit of $F_4$ as $s \to \infty$ should be bounded. (In general $f_0(s)$ is assumed to be bounded and the same for both case 1 and 2.) The solution of Eq. (42) is given in Appendix 2.

For Case 1, associated with smooth $v_1$ and $v_2$ with $f_0$ and $f_1$ given by Eq. (34) and $f_{a_i} = \tanh(s - s_0)$, the following solutions exist,

$$\frac{\mathrm{d}}{\mathrm{d}s} F_4(s) = \frac{c_4 \Phi(s)}{F_4^2(s)} \tag{44}$$

with solution,

$$F_4(s) = \sqrt[3]{3c_4 \int \Phi(s)\,ds + C_1} \tag{45}$$

If $\Phi(s) = -\lambda$ where $\lambda > 0$, then,

$$F_4 = \sqrt[3]{-c_4 \lambda s + C_1}$$

and a non-smooth solution in time $s$ is observed, that is a finite time blowup at $s = s_0$ for this $\Phi(s)$ starting with the first derivative of $F_4$ and higher. Define $\Phi(s)$ such that $F_4 = B + \sqrt[3]{-c_4 \lambda s + C_1}$, where $B$ is a constant. It can be verified that near $s = s_0$ Eq. (14) gives $\Phi(s) = -\lambda$ from the tensor product term. Furthermore the spatial solution is given by,

$$\begin{aligned}
&-4c_4F_5^3\left(y_1^2+y_2^2+y_3^2\right)\left(\frac{\partial F_5}{\partial y_1}-\frac{\partial F_5}{\partial y_2}\right)\frac{\partial^2 F_5}{\partial {y_3}^2}+2c_4F_5^2\left(\frac{\partial F_5}{\partial y_2}\right)\left(y_1^2+y_2^2+y_3^2\right)\left(\frac{\partial F_5}{\partial y_1}\right)^2\\
&+\Bigg(-2c_4F_5^2\left(y_1^2+y_2^2+y_3^2\right)\left(\frac{\partial F_5}{\partial y_2}\right)^2-4c_4F_5^3\left(y_1-y_2\right)\frac{\partial F_5}{\partial y_2}\\
&-\left(4y_1^2+4y_2^2+4y_3^2\right)\left(\frac{\partial F_5}{\partial y_3}\right)\left(c_4F_5^2\frac{\partial F_5}{\partial y_3}-1/4\right)\Bigg)\frac{\partial F_5}{\partial y_1}\\
&+\left(4y_1^2+4y_2^2+4y_3^2\right)\left(\frac{\partial F_5}{\partial y_3}\right)\left(c_4F_5^2\frac{\partial F_5}{\partial y_3}-1/4\right)\frac{\partial F_5}{\partial y_2}\\
&=0
\end{aligned} \tag{46}$$

In Appendix 2, the general solution for $F_5$ PDE associated with Case 1 is,

$$F_5(y_1,y_2,y_3) = G(y_1,y_2)\sqrt{B}\left[-m/2+3/2m\left(\csc\left(1/4\frac{2^{2/3}y_3}{\sqrt{-G(y_1,y_2)B\ 2^{2/3}}}+C/2\sqrt{6}\sqrt{m}\right)\right)^2\right]$$

Using the continuity equation a specific form for the surface $y_3 = F(y_1,y_2)$ emerges. Substitution of $F_1, F_2$ and $F_5$ (on the surface $y_3 = -F(y_1,y_2)$ into the continuity equation gives a surface to be described below, Differentiating $v_3$ wrt to $y_3$ gives,

$$\begin{aligned}
\frac{\partial v_3}{\partial y_3} = -3/4\frac{G(y_1,y_2)\sqrt{B}m^{3/2}2^{2/3}\sqrt{6}}{\sqrt{-BG(y_1,y_2)2^{2/3}}}\left(\csc\left(\left(1/4\frac{2^{2/3}y_3}{\sqrt{-BG(y_1,y_2)2^{2/3}}}+C/2\right)\sqrt{6}\sqrt{m}\right)\right)^2\times\\
\cot\left(\left(1/4\frac{2^{2/3}y_3}{\sqrt{-BG(y_1,y_2)2^{2/3}}}+C/2\right)\sqrt{6}\sqrt{m}\right)
\end{aligned}$$

Taking the limit as $m$ approaches 0 and consequently $C$ and $A$ approaching zero gives, where the continuity equation has been used to set $\frac{\partial v_3}{\partial y_3} = -\frac{\partial v_1}{\partial y_1} - \frac{\partial v_2}{\partial y_2}$,

$$4\,\frac{B^{3/2}\left(G(y_1,y_2)\right)^2\sqrt[3]{2}}{{y_3}^3} = -(2y_1+2y_2)\left({y_3}^2+A\right)(y_1y_2+A)$$

Solving algebraically for $y_3$ gives two roots and setting the result to $y_1^2 - y_2^2$, a saddle surface form, gives an equation which can be solved for $G(y_1,y_2)$, giving,

$$\frac{\sqrt[5]{2}\sqrt[5]{B^{3/2}\left(G(y_1,y_2)\right)^2\sqrt[3]{2}{y_1}^4{y_2}^4(y_1+y_2)^4}}{y_1y_2(y_1+y_2)} = {y_1}^2-{y_2}^2$$

$G$ is solved for and is exactly,

$$G(y_1,y_2) = \pm 1/2\,\frac{\sqrt[6]{2}\sqrt{\sqrt[3]{2}B^{3/2}y_1y_2(y_1-y_2)(y_1+y_2)^3(y_1-y_2)^2}}{B^{3/2}}$$

Substituting $G(y_1,y_2)$ into the expression for $F_5$ gives,

$$F_5 = \pm y_1y_2(y_1-y_2)^3(y_1+y_2)^4$$

with plot in **Figure 2a** and **b**.

Recalling the transformations to $\eta$ variables, we now return to star variables for the original PNS system. These are shown in **Figure 2d** in star variables. Note that the plot in **Figure 1c**, shows the range in $y$. In Eq. (8), the $\mathcal{G}_{\delta 4}$ term consists of the expression which has implicitly been set to zero,

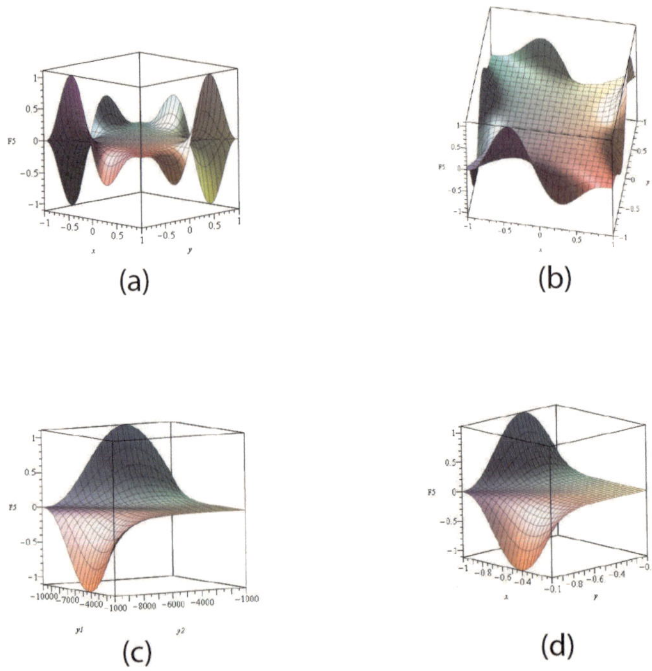


**Figure 2.**
*Plots of oscillations at boundary of cell in star and unstarred variables. (a) Sinusoidal velocity at boundary of cell for v3. (b) Different perspective for sinusoidal velocity at boundary of cell for v3. (c) Sinusoidal velocity at boundary of cell for v3 in terms of $y_i$ coordinates, $\frac{-y1y2(y1-y2)^3(y1+y2)^4}{3\times 10^{35}}$. (d) Sinusoidal velocity at boundary of cell for v3 in terms of $x_i$ coordinates, $\frac{-xy(10000x-10000y)^3(10000x+10000y)^4}{3\times 10^{35}}$.*

$$\Lambda_z = -\delta^3 v_3 \frac{\partial v_3}{\partial y_3} F_z + \delta^2 \vec{v} \cdot \nabla(v_3 F_z) = 0$$

Substituting $v_3 = c_4 F_5 (s - s_0)^{1/3}$ into this equation results in approximately zero as $m \to 0$ as $F_z$ is unbounded at corners and $v_3$ is zero at the corners. In **Figure 3c** $\kappa$ for $\vec{u} = u_x \vec{i} + u_y \vec{j} + u_z \vec{k}$ is shown after transforming from unstarred variables to starred ones. One can note that cancelation of oscillations will occur in a finite Lattice for $\kappa$ at the wall of adjacent cells since there the sinusoid is of equal height everywhere on $[-1, 1]$. Oscillations at infinity can occur as the Lattice dimension approaches infinity. A cusp-like bifurcation in vorticity field occurs indicating that there is a singularity upon using the correct definition of vorticity. This is shown in **Figure 3d**. Following the same recipe as above for case 1 $v_1, v_2$ smooth with $v_3$ oscillating at wall and a blowup in acceleration in time, case 2 has a new term now for $v_1 + v_2$ in the continuity equation. Using this new right hand side expression it follows after some calculations that a spatially singular $F_5$ is given as,

$$F_5 = \frac{-{y_1}^3 - {y_1}^2 y_2 + y_1 {y_2}^2 + {y_2}^3}{\left({y_1}^2 - {y_2}^2\right)^2 {y_1}^3 {y_2}^3}$$

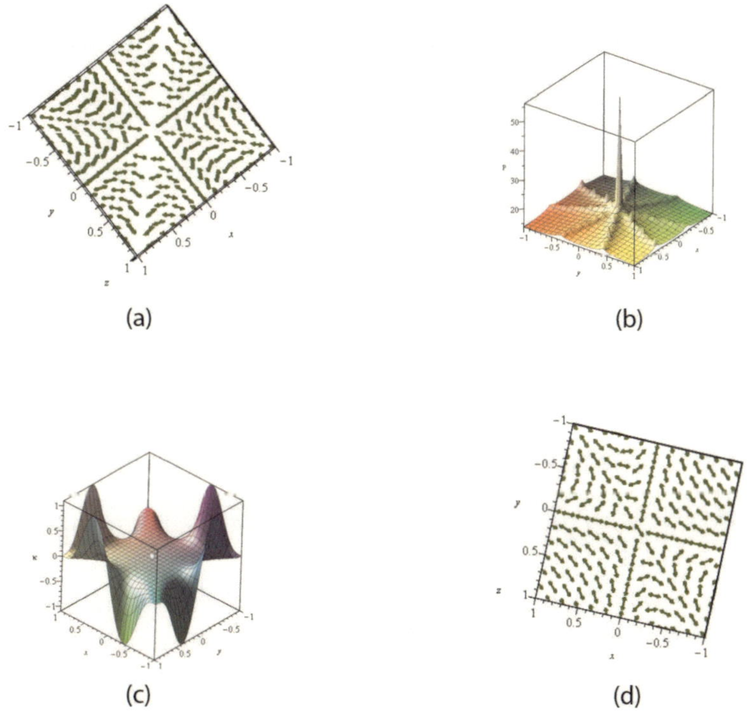


**Figure 3.**
*a. Plot of spatial blowup at center of cell for case 2, b. pressure function for case 1 and 2, c. Liutex magnitude difference $\kappa$ defined on the saddle surface $z = x^2 - y^2$, for temporally non-smooth $u_z$ case 1 and d. the existence of a vortex is found. (a) Spatial blowup at center of cell for no-finite time blowup. Excluding the origin along main principle axis cusp bifurcations are observed. (b) Pressure function in a given cell comprising of linear and nonlinear parts for both cases 1 and 2 for $u_z$. (c) Liutex magnitude difference $\kappa$. (defined on the saddle surface $z = x^2 - y^2$, for temporal non-smooth $u_z$). (d) Top view of the real vector field for $\omega = (\eta, \xi, R + \varepsilon)$ in terms of pure rotation, shearing and stretching. Here there is a cusp-like bifurcation along main principle axis in vorticity field. It was also evident that about y = x plane the field was parabolic. A non zero vorticity shows the existence of a vortex.*

with plot of vorticity in **Figure 3a**. Finally the solutions for $v_1, v_2$ and $v_3$ in case 1 can be verified to satisfy the continuity equation if and only if $m \to 0$ for any $\eta$ arbitrarily small and positive and for case 2 the solutions for $v_1, v_2$ and $v_3$ can be verified to satisfy to arbitrary small precision the continuity equation if and only if $m \to 0$ and $\eta$ is arbitrarily small and positive. Both of these hold on a saddle surface. Summarizing, for non smooth inputs $v_1$ and $v_2 = -v_1, \Phi(s) = 0$ and $f_0(s) \neq 0$ (a general function of $s$) gives $v_3$ a no finite time blowup in $s$, on the other hand for smooth inputs $v_1, v_2 = v_1$ and $f_0(s) = 1, \Phi(s) \neq 0$ (see Appendix 4) gives a finite time blowup in $s$ for the derivative of $v_3$ wrt to $s$. Solving for pressure for case 1 and then for case 2 velocities separately and thereby equating the second derivatives of pressure for each expression with respect to $y_3$ gives a new PDE for pressure. The plot is shown in **Figure 3b**. Here it can be seen that there is a max point and on the crests of the distribution function, the pressure is linear in $y_1$ and $y_2$. On the curved portions there will be finite time blowup starting with the first derivative wrt to $s$ as $\Phi(s) \neq 0$. The form of the solution for pressure $P$ associated with the non-blowup is $P = R(s)(Ay_1 + By_2 + C)$.

## 4. General solution with no restrictions on forcing and spatial velocities

Here there are no assumptions made on forcing and spatial velocities as being separable in space and time (**Figure 4**). PNS is made of the following component parts $C_i$ and is given by Eq. (47) (the same as Eq. (58) in Appendix 3- written in terms of special terms $\Phi_1$ here).

$$C_1 + C_2 = C_3 \tag{47}$$

$$\begin{aligned} C_1 = {} & \left(\frac{\partial v_3}{\partial s}\right)(v_3)^2\left(\frac{\partial v_3}{\partial y_1} - \frac{\partial v_3}{\partial y_2}\right)\frac{\partial^3 v_3}{\partial {y_3}^2 \partial s} - (v_3)^2\left(\frac{\partial v_3}{\partial y_1} - \frac{\partial v_3}{\partial y_2}\right)\left(\frac{\partial^2 v_3}{\partial y_3 \partial s}\right)^2 - \\ & \left(-2\left(\frac{\partial v_3}{\partial y_3}\right)v_3\frac{\partial v_3}{\partial s} + \Phi_1 + \Phi(s)\right)\left(\frac{\partial v_3}{\partial y_1} - \frac{\partial v_3}{\partial y_2}\right)\frac{\partial^2 v_3}{\partial y_3 \partial s} \end{aligned} \tag{48}$$

$$\begin{aligned} C_2 = {} & \left[\left(\frac{\partial v_3}{\partial y_1} - \frac{\partial v_3}{\partial y_2}\right)\frac{\partial \Phi_1}{\partial y_3} + 3\left(\frac{\partial v_3}{\partial s}\right)v_3\left(\frac{\partial v_3}{\partial y_1} - \frac{\partial v_3}{\partial y_2}\right)\frac{\partial^2 v_3}{\partial {y_3}^2}\right. \\ & - \left(2\frac{\partial v_3}{\partial y_1} - 2\frac{\partial v_3}{\partial y_2}\right)v_3\left(\frac{\partial v_3}{\partial y_1}\right)\frac{\partial^2 v_3}{\partial y_1 \partial s} - \left(2\frac{\partial v_3}{\partial x} - 2\frac{\partial v_3}{\partial y_2}\right)v_3\left(\frac{\partial v_3}{\partial y_2}\right)\frac{\partial^2 v_3}{\partial y_2 \partial s} \\ & + 2v_3\left(\frac{\partial v_3}{\partial y_1}\right)^2\left(\frac{\partial^2 v_3}{\partial y_1 \partial s} + \frac{y_3 N}{y_1^2 + y_2^2 + y_3^2}\right) - 2v_3\left(\frac{\partial v_3}{\partial y_2}\right)^2\left(\frac{\partial^2 v_3}{\partial y_2 \partial s} - \frac{\partial}{\partial s}\left(\frac{2(y_2 - y_1)v_3}{y_1^2 + y_2^2 + y_3^2}\right)\right) \\ & \left. + 3\left(\frac{\partial v_3}{\partial y_3}\right)^2\left(\frac{\partial v_3}{\partial s}\right)\left(\frac{\partial v_3}{\partial y_1} - \frac{\partial v_3}{\partial y_2}\right)\right]\frac{\partial v_3}{\partial s} \end{aligned} \tag{49}$$

$$\begin{aligned} C_3 = {} & \left[2\frac{(-2y_1 + 2y_2)\frac{\partial v_3}{\partial s} + 2y_3\left(\frac{\partial v_1}{\partial s} - \frac{\partial v_2}{\partial s}\right)}{{y_1}^2 + {y_2}^2 + {y_3}^2} - 2\frac{\partial^2 v_3}{\partial y_2 \partial s} + 2\frac{\partial^2 v_3}{\partial y_1 \partial s}\right] \times \\ & v_3\left(\frac{\partial v_3}{\partial y_1}\right)\left(\frac{\partial v_3}{\partial s}\right)\frac{\partial v_3}{\partial y_2} \end{aligned} \tag{50}$$

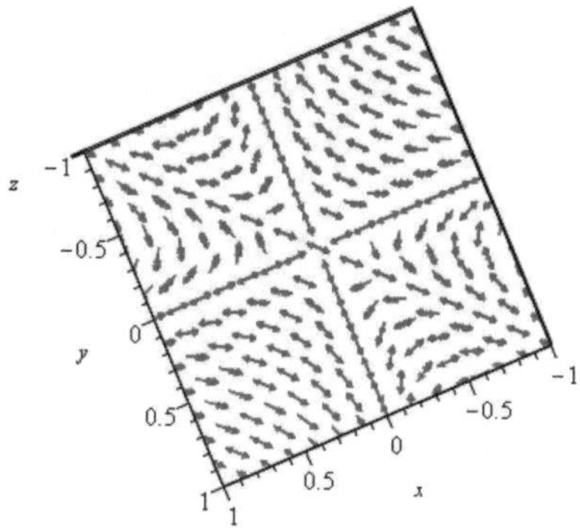


**Figure 4.**
*Top view of imaginary vector field-perpendicular along y = −x.*

where $\Phi_1$ is given as,

$$\Phi_1 = -\left(\frac{\partial v_3}{\partial y_3}\right) v_3 \frac{\partial v_3}{\partial s} - v_3^2 \frac{\partial^2 v_3}{\partial y_3 \partial s} - \frac{1}{2}\Phi(s) \tag{51}$$

It is immediate that Eq. (47) gives as solution the same form as in the main section of this paper, that is the solution of Eq. (45) multiplied by $F_5(y_1, y_2, y_3)$. The expression for $\Phi_1$ is determined by solving for $\vec{F} \cdot \nabla v_3^2 + \nabla v_3^2 \cdot \frac{\partial \vec{b}}{\partial s}$ in Eq. (4) when $\delta \approx 1$. Simplifying Eq. (47) using Eq. (51) and using the definition of the Liutex part of vorticity in Eqs. (27) and (28) and solving algebraically for $N = \frac{\partial v_1}{\partial s} - \frac{\partial v_2}{\partial s}$ in Eq. (48) the problem is simplified to,

$$-\frac{\left(\frac{\partial v_3}{\partial y_1} - \frac{\partial v_3}{\partial y_2}\right)\left(-1/4\Phi(s)\left({y_1}^2 + {y_2}^2 + {y_3}^2\right)\frac{\partial^2 v_3}{\partial y_3 \partial s} + \left(\frac{\partial v_3}{\partial s}\right)^2\left(v_3\left({y_1}^2 + {y_2}^2 + {y_3}^2\right)\frac{\partial^2 v_3}{\partial {y_3}^2} + 2v_3(-y_2 + y_1)\frac{\partial v_3}{\partial y_2} + \left(\frac{\partial v_3}{\partial y_3}\right)^2\left({y_1}^2 + {y_2}^2 + {y_3}^2\right)\right)\right)}{y_3\left(\frac{\partial v_3}{\partial s}\right)\left(\frac{\partial v_3}{\partial y_1}\right)v_3\left(\frac{\partial v_3}{\partial y_1} - 2\frac{\partial v_3}{\partial y_2}\right)} - \frac{\partial v_1}{\partial s} + \frac{\partial v_2}{\partial s} = 0 \tag{52}$$

This PDE is separable as $v_3 = F_7(y_1, y_2, y_3)F_4(s)$, where,

$$4F_7^3\left(\frac{\partial^2 F_7}{\partial {y_3}^2}\right){y_1}^2 + 4F_7^3\left(\frac{\partial^2 F_7}{\partial {y_3}^2}\right){y_2}^2 + 4F_7^3\left(\frac{\partial^2 F_7}{\partial {y_3}^2}\right){y_3}^2 + 4F_7^2\left(\frac{\partial F_7}{\partial y_3}\right)^2 {y_1}^2 + 4F_7^2\left(\frac{\partial F_7}{\partial y_3}\right)^2 {y_2}^2 + 4F_7^2\left(\frac{\partial F_7}{\partial y_3}\right)^2 {y_3}^2 + 8F_7^3\left(\frac{\partial F_7}{\partial y_2}\right)y_1 - 8\left(F_7^3\right)\left(\frac{\partial F_7}{\partial y_2}\right)y_2 - \left(\frac{\partial F_7}{\partial y_3}\right){y_1}^2 c_4 - \left(\frac{\partial F_7}{\partial y_3}\right){y_2}^2 c_4 - \left(\frac{\partial F_7}{\partial y_3}\right){y_3}^2 c_4 = 0$$

We have used the following equalities,

$$\frac{\partial^2 v_1}{\partial y_3 \partial s} = \frac{\partial^2 v_3}{\partial y_1 \partial s} + \frac{y_3 N}{y_1^2 + y_2^2 + y_3^2}$$

$$\frac{\partial^2 v_2}{\partial y_3 \partial s} = \frac{\partial^2 v_3}{\partial y_2 \partial s} - \frac{\partial}{\partial s}\left(\frac{2(y_2 - y_1)v_3}{y_1^2 + y_2^2 + y_3^2}\right)$$

where $N$ is defined by,

$$N = \frac{\partial v_1}{\partial s} - \frac{\partial v_2}{\partial s}$$

## 5. Discussion and conclusion

Comparing solutions for Eqs. (42)–(46) it is evident that the former one has as solution for $u_z$ a time component that is smooth(except at infinity) opposite to that of the smooth force and spatially non-smooth $y_1$ and $y_2$ velocities whereas in the latter equations $u_z$ has a finite time blowup(with the first derivative and higher) for $f_0$,$f_1$ and $y_1$ and $y_2$ smooth velocity function inputs. (see Eq. (45)) [4]. Eq. (47) is the full non-separable reduced PDE. Oscillations of arbitrary height can occur at spatial infinity. For (PNS) it is shown that there exists a vortex in each cell of the lattice associated with $\mathbb{T}^3$ using the decomposition of pure rotation(Liutex), antisymmetric shear and compression and stretching. A cusp bifurcation for vorticity shows the birth and destruction of vorticies. Here it is known that streaklines can be used to give an idea of where the vorticity in a flow resides. The question of no-finite time blowup for the new eqs. [1] replacing the Navier Stokes equations is left for future study.

## A. Appendix 1

Eq. (1) including forcing terms and the three associated velocities $v_1, v_2$ and $v_3$,

$$L = L_1 + 2(L_2 + L_3)\frac{\partial v_3}{\partial s} = 0 \tag{53}$$

$$\begin{aligned} L_1 = {} & v_3^3\left(\frac{\partial v_3}{\partial s}\right)^2 \mu(-1+\delta)\frac{\partial^3 v_3}{\partial y_3 \partial y_1{}^2} + (v_3)^3\left(\frac{\partial v_3}{\partial s}\right)^2 \times \\ & \mu(-1+\delta)\frac{\partial^3 v_3}{\partial y_3 \partial y_2{}^2} + (v_3)^3\left(\frac{\partial v_3}{\partial s}\right)^2 \mu(-1+\delta)\frac{\partial^3 v_3}{\partial y_3{}^3} + \\ & (v_3)^5\left(\frac{\partial v_3}{\partial s}\right)\left(\frac{\partial^3 v_3}{\partial y_3{}^2 \partial s}\right)\delta\rho - \left(\frac{\partial^2 v_3}{\partial y_3 \partial s}\right)^2 (v_3)^5 \delta\rho - \\ & \left(-2(v_3)^3\left(\frac{\partial v_3}{\partial t}\right)\left(\frac{\partial v_3}{\partial y_3}\right)\delta + 2(v_3)^3\delta\left(F_{T1} + \frac{\partial v_1}{\partial s}\right)\frac{\partial v_3}{\partial y_1} + \right. \\ & \left. 2(v_3)^3\delta\left(F_{T2} + \frac{\partial v_2}{\partial s}\right)\frac{\partial v_3}{\partial y_2} - \delta + 1 + \frac{1}{2}v_3^2\Phi(s)\right)v_3\rho\frac{\partial^2 v_3}{\partial y_3 \partial s} \end{aligned} \tag{54}$$

where $\Phi$ is given by Eq. (19).

$$L_2 = \left[\frac{1}{2}(-1+\delta)(\delta v_1 - 1)\frac{\partial v_3}{\partial s} + v_3\delta\rho\left(F_{T1} + \frac{\partial v_1}{\partial s}\right)\right]v_3^3\frac{\partial^2 v_3}{\partial y_3 \partial y_1} +$$

$$v_3^3\left(\frac{1}{2}(-1+\delta)(\delta v_2 - 1)\frac{\partial v_3}{\partial s} + v_3\delta\rho\left(F_{T2} + \frac{\partial v_2}{\partial s}\right)\right)\frac{\partial^2 v_3}{\partial y_3 \partial y_2} + \frac{1}{2}\left(-2+3\left(\rho+\frac{2}{3}\right)\delta\right)\left(\frac{\partial v_3}{\partial s}\right)$$

$$v_3^4\frac{\partial^2 v_3}{\partial {y_3}^2} + v_3^4\left(\frac{\partial v_3}{\partial t}\right)(-1+\delta)\frac{\partial^2 v_3}{\partial {y_1}^2} + v_3^4\left(\frac{\partial v_3}{\partial s}\right)(-1+\delta)\frac{\partial^2 v_3}{\partial {y_2}^2} + v_3^4\left(\frac{\partial^2 v_1}{\partial y_3 \partial s}\right)\left(\frac{\partial v_3}{\partial y_1}\right)\delta\rho \tag{55}$$

$$L_3 = v_3^4\left(\frac{\partial^2 v_2}{\partial y_3 \partial s}v_2\right)\left(\frac{\partial v_3}{\partial y_2}\right)\delta\rho + \frac{1}{2}\left[(-1+(3\rho+1)\delta)\left(\frac{\partial v_3}{\partial y_3}\right)^2 + (-1+\delta)\left(\left(\frac{\partial v_1}{\partial y_1}\right)^2\right.\right.$$

$$+2\left(\frac{\partial v_1}{\partial y_2}\right)\frac{\partial v_2}{\partial y_1} + \left(\frac{\partial v_2}{\partial y_2}\right)^2)](v_3)^3\frac{\partial v_3}{\partial s} + \left(\left((v_3)^3\delta\left(F_{T1} + \frac{\partial v_1}{\partial s}\right)\frac{\partial v_3}{\partial y_1} +\right.\right.$$

$$(v_3)^3\delta\left(F_{T2} + \frac{\partial v_2}{\partial s}\right)\frac{\partial v_3}{\partial y_2} - 1 + \delta).\frac{\partial v_3}{\partial y_3} + (v_3)^4\delta\left(\left(\frac{\partial v_3}{\partial y_1}\right)\frac{\partial F_{T1}}{\partial y_3} + \left(\frac{\partial v_3}{\partial y_2}\right)\frac{\partial F_{T2}}{\partial y_3}\right)\Big)\rho \tag{56}$$

## B. Appendix 2

Solving Eq. (52) by assuming,

$$F_5(y_1, y_2, y_3) = c_4 P\left(\frac{2^{2/3} y_3}{2\sqrt{-BG(y_1, y_2)2^{2/3}}} + C, 0, 0\right)G(y_1, y_2)\,\mathrm{sech}\,(F(y_1, y_2) + y_3)^2 \tag{57}$$

gives a product of two factors of equations, one of which has solution on an arbitrary surface $y_3 = -F(y_1, y_2)$ and the other on an $\epsilon$ ball containing this arbitrary surface.

Substituting Eq. (57) into Eq. (52) and solving algebraically for the second derivative of $v_3$ wrt to $y_3$ and the result was set equal to $v_3^2$. The resulting equation is checked to see if it has a Weierstrass P function as solution, and it does by solving for it using the *pdsolve* command. In applying the Geometric Algebra method used in [6] which this work is based on, it was checked that the term $\|\nabla u_z\|_2$ approaches zero and is an element of the Schwartz class for any constant $C$ and in particular for arbitrarily large values of $y_3$. It was assumed that $v_3$ is in the form $v_3 = F_4(s)G(y_1, y_2)U(y_3, s)$. $\Bigg($Also $v_3 = \frac{U(y_3, s)F(s)}{\sqrt[3]{s-s_0}\sqrt{G(y_1, y_2)}}$ solves the problem in question. Note the modular form $U(y_3, s) = \frac{2^{2/3}}{\sqrt[6]{G(y_1, y_2)}}P\left(1/2\frac{2^{2/3}y_3}{\sqrt{-(G(y_1, y_2))^{2/3}2^{2/3}}} + F_1(s), 0, 0\right)\Bigg)$

## C. Appendix 3

The following equation is similar to Eq. (29) with the exception that it is expressed more generally in terms of $v_1$ and $v_2$,

$$B_1 + 2\frac{\partial v_3}{\partial s}(B_2 + B_3) = B_4 \tag{58}$$

$$B_1 = \left(\frac{\partial v_3}{\partial s}\right)(v_3)^2\left(\frac{\partial v_3}{\partial y_1} - \frac{\partial v_3}{\partial y_2}\right)\frac{\partial^3 v_3}{\partial {y_3}^2 \partial s} - (v_3)^2\left(\frac{\partial v_3}{\partial y_1} - \frac{\partial v_3}{\partial y_2}\right)\left(\frac{\partial^2 v_3}{\partial y_3 \partial s}\right)^2 - \left(2\frac{\partial v_3}{\partial y_1} - 2\frac{\partial v_3}{\partial y_2}\right) \times \left[-\left(\frac{\partial v_3}{\partial y_3}\right)v_3\frac{\partial v_3}{\partial s} + v_3\left(F_{T_1} + \frac{\partial v_1}{\partial s}\right)\frac{\partial v_3}{\partial y_1}v_3 + v_3\left(F_{T_2} + \frac{\partial v_2}{\partial s}\right)\frac{\partial v_3}{\partial y_2} + \frac{1}{2}\Phi(t)\right]\frac{\partial^2 v_3}{\partial y_3 \partial s} \tag{59}$$

$$B_2 = v_3\left(\frac{\partial v_3}{\partial y_1} - \frac{\partial v_3}{\partial y_2}\right)\left(F_{T_1} + \frac{\partial v_1}{\partial s}\right)\frac{\partial^2 v_3}{\partial y_3 \partial y_1} + v_3\left(\frac{\partial v_3}{\partial y_1} - \frac{\partial v_3}{\partial y_2}\right)\left(F_{T_2} + \frac{\partial v_2}{\partial t}\right)\frac{\partial^2 v_3}{\partial y_3 \partial y_2} + \frac{3}{2}\left(\frac{\partial v_3}{\partial s}\right)v_3\left(\frac{\partial v_3}{\partial y_1} - \frac{\partial v_3}{\partial y_2}\right)\frac{\partial^2 v_3}{\partial {y_3}^2} \tag{60}$$

$$B_3 = \left(\frac{\partial v_3}{\partial y_1}\right)^2\left(\frac{\partial^2 v_1}{\partial y_3 \partial s}\right)v_3 - \left(\frac{\partial v_3}{\partial y_2}\right)^2\left(\frac{\partial^2 v_2}{\partial y_3 \partial s}\right)v_3 + \left[\frac{3}{2}\left(\frac{\partial v_3}{\partial y_3}\right)^2\frac{\partial v_3}{\partial s} + \left(\left(F_{T_1} + \frac{\partial v_1}{\partial s}\right)\frac{\partial v_3}{\partial y_3} + v_3\frac{\partial F_{T_1}}{\partial y_3}\right)\frac{\partial v_3}{\partial y_1} + \left(\left(F_{T_2} + \frac{\partial v_2}{\partial s}\right)\frac{\partial v_3}{\partial y_3} + v_3\frac{\partial F_{T_2}}{\partial y_3}\right)\frac{\partial v_3}{\partial y_2}\right]\left(\frac{\partial v_3}{\partial y_1} - \frac{\partial v_3}{\partial y_2}\right) \tag{61}$$

$$B_4 = \left[2\frac{(-2y_1 + 2y_2)\frac{\partial v_3}{\partial s} + 2y_3\left(\frac{\partial v_1}{\partial s} - \frac{\partial v_2}{\partial s}\right)}{{y_1}^2 + {y_2}^2 + {y_3}^2} - 2\frac{\partial^2 v_3}{\partial y_2 \partial s} + 2\frac{\partial^2 v_3}{\partial y_1 \partial s}\right]v_3\left(\frac{\partial v_3}{\partial y_1}\right)\left(\frac{\partial v_3}{\partial s}\right)\frac{\partial v_3}{\partial y_2} \tag{62}$$

## D. Appendix 4

The second term in Eq. (14) involving the tensor product expression for the solution of $v_3$ given in terms of $F_5$ is independent of s $(\Phi(s) = H_2(y_1, y_2, y_3))$, and is given as,

$$1/3\frac{(G_2(y_1,y_2,y_3))^2\left(\sqrt[3]{s-s_0}\frac{\partial}{\partial y_1}G_2(y_1,y_2,y_3) + \sqrt[3]{s-s_0}\frac{\partial}{\partial y_2}G_2(y_1,y_2,y_3) + \sqrt[3]{s-s_0}\frac{\partial}{\partial y_3}G_2(y_1,y_2,y_3)\right)}{\sqrt[3]{s-s_0}}$$
$$= 1/3(G_2(y_1,y_2,y_3))^2\left(\frac{\partial}{\partial y_1}G_2(y_1,y_2,y_3) + \frac{\partial}{\partial y_2}G_2(y_1,y_2,y_3) + \frac{\partial}{\partial y_3}G_2(y_1,y_2,y_3)\right) \tag{63}$$

The surface integral in Eq. (14) involving the pressure terms $Q$ is either zero or non-zero depending on if $\Phi(s) = 0$ which occurs when $v_1 = -v_2$ or $\Phi(s) \neq 0$ which occurs when $v_1 = v_2$. In general we have the term $K_1 - \Phi(s) = \lambda_1$, where $\lambda_1 \neq 0$. The term $K_1 - \Phi(s)$ is associated with taking the gradient of the extended expression $\nabla_D^{-1}(\mathcal{G}_{\delta 1} + \mathcal{G}_{\delta 2} + \mathcal{G}_{\delta 3} + \mathcal{G}_{\delta 4}) = \vec{r}$.

## Author details

Terry E. Moschandreou[1*] and Keith C. Afas[2,3]

1 Thames Valley District School Board, London, ON, Canada

2 School of Biomedical Engineering, University of Western Ontario, London, ON, Canada

3 Department of Medical Biophysics, University of Western Ontario, London, ON, Canada

*Address all correspondence to: tmoschandreou@gmail.com

## References

[1] Liu C, Liu Z. New governing equations for fluid dynamics. AIP Advances. 2021;**11**:115025. DOI: 10.1063/5.0074615

[2] Benzi R, Colella M. A simple point vortex model for two-dimensional decaying turbulence. Physics of Fluids A: Fluid Dynamics. 1991;**4**(5):1-11. DOI: 10.1063/1.858254

[3] Santangelo P. The generation of vortices in high-resolution, two-dimensional decaying turbulence and the influence of initial conditions on the breaking of self-similarity. Physics of Fluids A:Fluid Dynamics. 1989;**1**(6): 1027-1034. DOI: 10.1063/1.857393

[4] T.E. Moschandreou, Navier Stokes equations for a 3D incompressible fluid on $\mathbb{T}^3$ Submitted to Partial Differential equations in Applied Mathematics (June 10,2022)-currently under review. 2022. pp. 1-22

[5] Leray J, Terrell R. On the motion of a viscous liquid filling space. arXiv: 1604.02484 [math.HO]. 2016;**1**:1-113

[6] Moschandreou TE. No finite time blowup for 3D incompressible Navier stokes equations via scaling invariance. Mathematics and Statistics. 2021;**9**(3): 386-393

[7] Moschandreou TE. On the 4th clay millennium problem for the periodic Navier stokes equations millennium prize problems. Recent Advances in Mathematical Research and Computer Science. 2021;**4**(12):79-92. DOI: 10.9734/bpi/ramrcs/v4/14379D

[8] Moschandreou TE, Afas KC. Existence of incompressible vortex-class phenomena and variational formulation of Rayleigh Plesset cavitation dynamics. Applied Mechanics. 2021;**2**(3):613-629. DOI: 10.3390/applmech2030035

[9] DeTurck D, Gluck H, Komendarczyk R, Melvin P, Shonkwiler C, Vela-Vick DS. Triple linking numbers, ambiguous Hopf invariants and integral formulas for three-component links. Matemática Contemporânea. Sociedade Brasileira de Matemática. 2008;**34**:251-283

[10] Benguria RD, Vallejos C, Van Den BH. Gagliardo-Nirenberg-Sobolev inequalites for convex domains in $\mathbb{R}^D$. Arxiv.org arXiv:1802.01740 [math-ph]. 2018;**2**:1-16

[11] Temam R. Behaviour at time $t = 0$ of the solutions of semi-linear evolution equations. Journal of Differential Equations. 1982;**43**:73-92

[12] Temam R. Suitable initial conditions. Journal of Computational Physics. 2006; **218**:443-450

Chapter 2

# Mass Flux in Corkscrew Flow Vortices in the Venus Plasma Wake

*Hector Pérez-de-Tejada and Rickard Lundin*


## Abstract

Measurements conducted with the Venus Express spacecraft (VEX) around Venus have provided evidence for the presence of a vortex structure in its wake. A configuration of the form of a corkscrew flow with a cross-section comparable to the planet's radius has been inferred from those measurements and exhibits a rotation in the counterclockwise sense when viewed from the wake back to Venus. Such structure is generated by the solar wind and also by planetary ions driven along the wake as inferred from the analysis of data obtained in several orbits of that spacecraft. It has also been learned that the width of the corkscrew structure gradually decreases with distance along the wake and its position varies along the solar cycle occurring closer to the planet during minimum solar cycle conditions. Measurements also show that the flow speed of the planetary ions driven from the nightside ionosphere is modified as they move through the corkscrew flow structure and become accelerated as the width of a corkscrew structure decreases with increasing distance downstream from Venus. Measurements also show that the mass flux of the planetary ions increases at high altitudes above the planet when they are conducted across the narrow part of a corkscrew shape in the particle distribution along the wake.


**Keywords:** plasma vortex in the Venus wake, mass flux in the Venus wake, particle acceleration in the Venus wake, mass flux conservation, momentum exchange in the Venus wake

## 1. Introduction

Among the various features inferred from measurements conducted with the Pioneer-Venus (PVO) and the Venus Express (VEX) spacecraft in orbit around Venus there has been evidence of a vortex structure present along its wake [1–3]. As a whole its width in the near wake is comparable to the Venus radius and it is seen to exhibit a rotation in the counterclockwise direction when viewed from the wake [4]. In addition, the position of the vortex structure varies along the solar cycle and becomes more closely located to Venus under minimum solar cycle conditions [5, 6].

An implication of this latter behavior is that the gradual decrease of the vortex width with the downstream distance from Venus implies the acceleration of the planetary ions that stream in vortex structures that have smaller widths together with those that move along the wake direction. This effect is produced by the enhanced values of the kinetic energy of planetary ions that are forced to move in smaller-width vortex structures since it is required that the integrated energy flow value across the wake is preserved [7]. As a whole, the process is produced by the expansion of the solar wind plasma into the wake together with the planetary ions that are eroded from the ionosphere as they move over its magnetic polar regions [8].

In general terms, the momentum flux of the solar wind is gradually transferred to Venus upper ionosphere through viscous processes as it moves to the magnetic polar regions by the dayside [9]. As a result of the momentum transport through a velocity boundary layer adjacent to the ionopause a fraction of the available energy is dissipated and then is used to increase the local plasma temperature [10–12], Phillips and McComas [13]. Under such conditions, there is a forced entry of the plasma into the wake [8] which in turn gradually decreases with distance downstream from Venus thus decreasing the width of the ionospheric vortices. In this scheme vortex structures produced across the wake are subject to being restricted to move within an ever-decreasing region in the central wake thus leading to enhance the kinetic energy of the particles that stream in that region to maintain the integrated energy of vortex motion across the wake [7].

Much of that activity should ultimately result from wave-particle interactions between the local plasma populations and turbulent and fluctuating oscillations of the magnetic field convected by the solar wind and that have been measured in the Venus plasma environment [14, 15]. In that view, electrostatic and proton cyclotron waves [16–19] may be responsible for modifying the predicted large-scale trajectories of the planetary ions along the Venus plasma wake [20]. At the same time, those processes could influence the transport of solar wind momentum and its dissipation to account for the measured plasma heating in that region.

## 2. The VEX data

From measurements conducted with the ASPERA instrument in the VEX spacecraft, there are density and speed profiles of the solar wind and the planetary ions with altitude above the planet that indicate the way those variables vary along the dusk-dawn meridian plane and also in the noon-midnight plane. These are reproduced in **Figures 1** and **2** to show notable changes that occur at different altitudes. Even though the density values in the dawn-dusk plane (left panel in **Figure 1**) maintain decreasing values (down to ~20 $cm^{-3}$) at altitudes higher than the ionopause boundary labeled IP (by ~1000 km) there is a notable difference in the noon-midnight plane (left panel in **Figure 2**) where more intense density values are measured in the density profile above ~5000 km. This is particularly the case between 8000 and 10,000 km where much higher density values (reaching up to ~ 80 $cm^{-3}$) were recorded. At the same time, the speed profiles of the $H+$ and the planetary $O+$ ions first show a gradual similar variation with height above the ionopause in the right panel of both figures and that reaches a (~10 km/s) value by 3000–5000 km, thus implying a velocity shear above that boundary. At higher altitudes, the speed values in both figures begin an unexpected increase to higher (~ 40 km/s) values by ~10,000 km. This later variation is seen in the right panel of both figures and reveals the presence of different phenomena.

A possible interpretation of these changes can be advanced by considering that the Venus ionospheric plasma that is eroded by the solar wind from the nightside moves

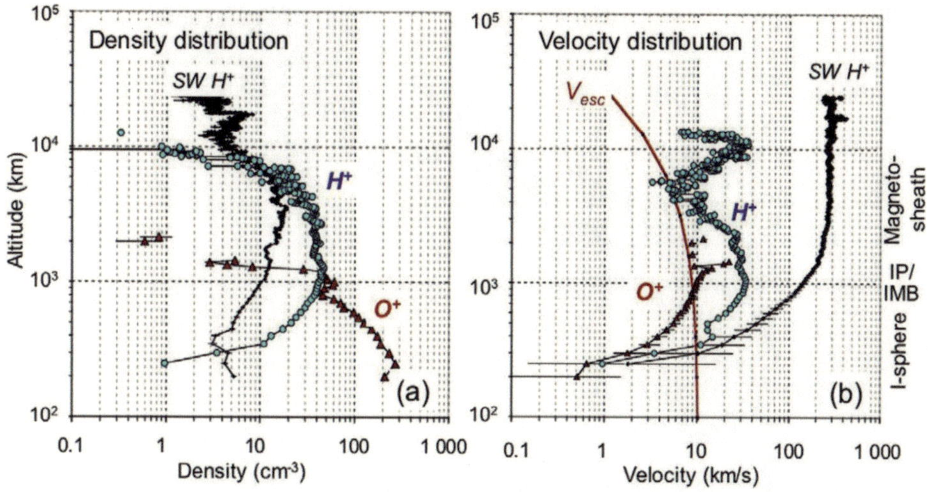


**Figure 1.**
*Dawn–dusk Meridian. Number densities (a) and flow velocities (b) versus spacecraft altitude for solar wind $H^+$ ions, and also for ionospheric $H^+$ and $O^+$ ions. Curve marked $V_{esc}$ illustrates escape velocity versus altitude above Venus. The data points represent average values in 50 km altitude intervals, sampled within Y = ±0.5 $R_V$ of the Dawn–dusk meridian. Error bars give the accuracy of individual measurement points. Regions and boundaries encountered are marked out on the right-hand side as: I-sphere (the core ionosphere), the Ionopause (IP/IMB), and the magneto-sheath (from [1]).*

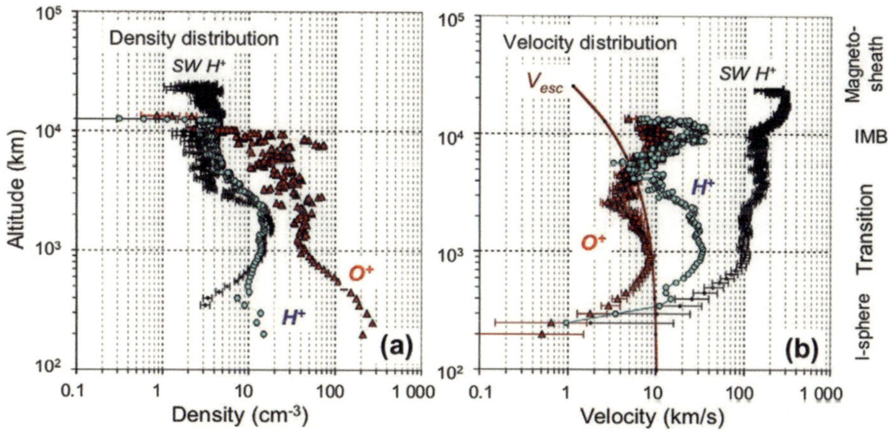


**Figure 2.**
*Noon-midnight Meridian. Density and speed values of the solar wind and ionospheric ions as a function of spacecraft altitude as in* **Figure 1** *measured in the noon-midnight meridian (from [1]).*

through a corkscrew geometry as illustrated in **Figure 3** and that corresponds to conditions like those in a similar fluid dynamic configuration. In the solar wind that streams past Venus and that is pressed into the wake behind the planet by thermal pressure forces [8, 21], there are plasma vortices whose width gradually decreases with distance along the wake [5]. In this scheme, particles that stream in the central part of the wake become accelerated by energy accumulated from vortices whose width gradually becomes reduced along that direction [7].

It is to be noted that the speed enhancement by ~10,000 km altitude in **Figures 1** and **2** is more intense in the noon-midnight meridian and that at even higher altitudes

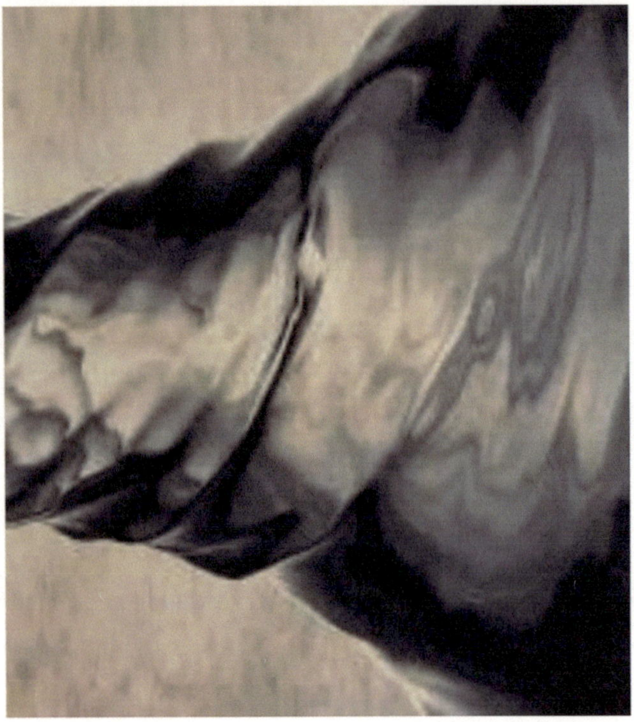

**Figure 3.**
*Corkscrew flow configuration. View of a corkscrew vortex flow in fluid dynamics. Its geometry is equivalent to that expected for a vortex flow in the Venus wake with its width and position varying during the solar cycle. The vortex flow becomes thinner along the radial direction in the wake axis when measured on a plane transverse to that of the figure further downstream from Venus [5].*

there is also a sudden increase in the speed profile of the solar wind ions. In general terms, it can be stated that the conditions encountered by the noon-midnight meridian are more intense than those measured by the dawn-dusk meridian and it is possible that they become accumulated from the flow driven along the sides of the planet and they are led to the noon-midnight region of the wake.

## 3. The corkscrew flow

There are two main properties in the density and speed profiles in the panels of **Figures 1** and **2**. While the density values in the left panels of both figures follow a nearly smooth decrease with altitude above the ionopause at ~$10^3$ km (most notable in the dawn-dusk meridian figure), there are notable density variations above the 5000 km altitude in the noon-midnight distribution. At that altitude, there is also a sudden change in the speed profiles in the right panel of both figures with increasing values that reach ~40 km/s by 10,000 km. The latter change in the altitude gradient is peculiar since it points out an effect that occurs as a result of a different fluid dynamic configuration. The geometry presented in **Figure 3** can account for such change since it may derive from the spacecraft first moving away from the main ionospheric body and crossing later the narrow shape of a corkscrew flow configuration.

For simplicity let us assume that **Figure 3** is on the noon-midnight meridian and it represents the inner part of its wake which extends to the far-left side. At the same time, the VEX spacecraft follows a linear trajectory that leaves the ionosphere from the north-right side of the figure moves then nearly above its boundary and later enters and crosses the thinner part of the corkscrew flow which is formed by ionospheric plasma that has been eroded by the solar wind. Since we cannot make a quantitative estimate of the thickness of the later region nor the inclination of the spacecraft trajectory it is only possible to provide a schematic description of the proposed model.

## 4. Mass flux profiles

The deformed shape of the Venus nightside ionosphere implied from **Figure 3** should be acquired as a result of the erosion produced by the solar wind that streams over its polar regions leading to the corkscrew flow geometry. Even though there is no direct information available on the thickness and extent of the central wake it is of interest to consider that its crossing can be assumed to be a separate event in the ionospheric density and speed profiles measured by the VEX spacecraft as it moves through that region. Such an interpretation can be made from **Figures 1** and **2** since the enhanced density and speed values measured in the 8000–10,000 km altitude range are unrelated to those in the lower altitude nightside ionosphere. Instead, they occur under different conditions since at high altitudes they disrupt the gradual and persistent density decrease in their profiles and also include high values in the velocity profiles more clearly shown in the right panel of **Figure 2**.

As a result, the unexpected high speed and density values at high altitudes can be viewed as crossing the narrow part of the corkscrew flow that extends downstream in the central Venus wake. The purpose of that feature can be explored by examining changes that are implied on the density and speed of the flow when VEX is subject to stream within the thin part of the corkscrew flow. In fact, since the cross-section of the flow vortex decreases with distance along the wake the area integrated value of the mass flux should be confined within a region with smaller cross-section along the wake thus implying higher speed and density values that should be encountered.

The results provided by the data are presented in **Figure 4** where values of the mass flux derived from the profiles presented in **Figures 1** and **2** describe its distribution with altitude for the dawn-dusk meridian (left panel) and to the

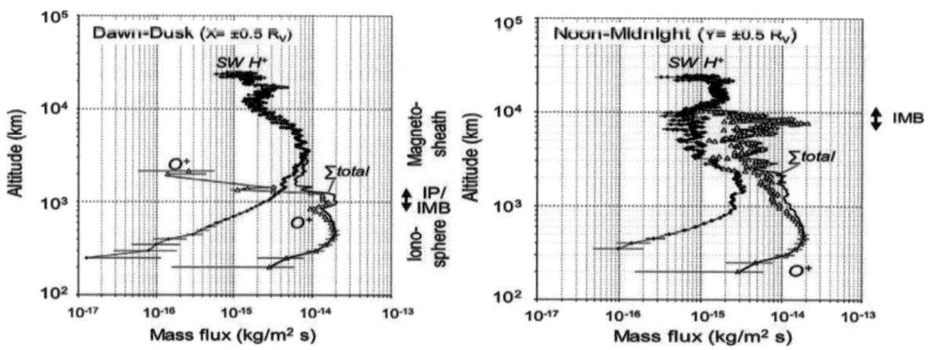


**Figure 4.**
*Mass flux profiles. (left) average solar wind $H^+$, ionospheric $O^+$ (triangles), and total mass flux values on the dawn–dusk meridian (left) and to the noon-midnight meridian (right). The data are averaged in 50 km altitude intervals. The average position of the Ionopause and the induced magnetosphere boundary is marked by IP and IMB (from [1]).*

noon-midnight meridian (right panel). The dominant feature is that superimposed on the values corresponding to an undisturbed density data distribution as that obtained in the dawn-dusk meridian there is a notable increase of the mass flux by the 8000–10,000 altitude range where in the right panel it reaches (~ $2.10^{-14}$ kg/ $m^2$ s) values at high altitudes and that are twice as large as those in the left panel (~ $6.10^{-15}$ kg/$m^2$s).

## 5. Calculations

A detailed evaluation of the corresponding mass flux values expected from the density and speed altitude profiles reproduced in **Figure 4** can be obtained by estimating changes in the cross-section value of the flow as it moves along the wake. A useful view is available from the distribution of the length of the vortices measured across the Venus wake during the VEX years of operation between 2006 and 2013. This is shown in two separate circles in **Figure 5** representing conditions measured before the minimum solar cycle between 2006 and 2009 (left circle) and those that occurred during and after those conditions between 2010 and 2013 (right circle). It is notable that segments are placed at a position on the vertical axis that corresponds to the time duration of the vortex between the entry and the exit of the spacecraft and are located at a higher value

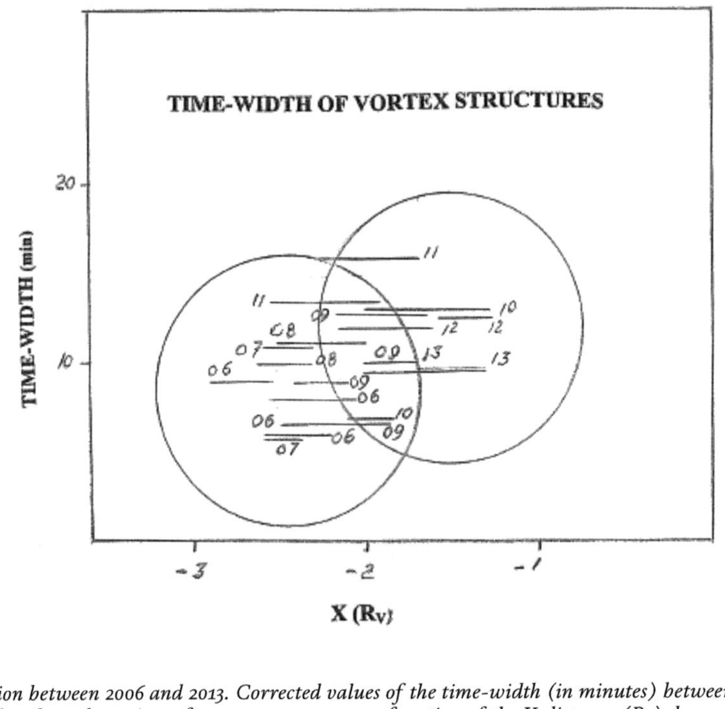


**Figure 5.**
*Vortex position between 2006 and 2013. Corrected values of the time-width (in minutes) between the VEX inbound and outbound crossings of vortex structures as a function of the X-distance ($R_V$) downstream from Venus that were measured in 20 orbits. The numbers at the side of each segment represent the two last digits of the year when measurements were made in different orbits between 2006 and 2013 (four orbits were examined during 2006 and also during 2009). The two circles confine orbits between 2006 and 2009 (left circle) prior to a solar cycle minimum and those between 2009 and 2013 (right circle) during and after that period [5].*

in the right circle than those in the left circle. At the same time, they extend as well through smaller X-distances (along the Sun-Venus line) in the right circle than those in the left circle (downstream along the tail). Such differences imply the varying position and width of the vortices with thicker values in the right circle that occur closer to Venus. This general characteristic of the vortices is the same as that described in **Figure 3** for a corkscrew flow where their width is wider on the right-hand side.

A case-by-case analysis of this peculiarity was conducted by calculating the average values of both variables (position and width) for each orbit during the VEX entry and exit across the vortices. The results are shown in the two circles of **Figure 5** with numbers indicating the two last digits of the year of each measurement. Most remarkable is that smaller $\Delta T$ width values were obtained for the 2006–2009 orbits (left circle) corresponding to vortices in the thin section of the corkscrew flow configuration in **Figure 3**. This agreement provides evidence for a similar description of plasma motion in the Venus wake. As a whole, it can be stated that there is a general downward displacement of the segment position between those in the 2010–2013 orbits (right circle) and those in the 2006–2009 orbits (left circle). In fact, the segments in the latter case occur by $X = -2.5\ R_V$ in the vicinity of $\Delta T$ ~7 min while those for the former orbits (right circle) are placed in the $X \sim -1.5\ R_V$ region with a larger ($\Delta T$ ~12 min) time span. As a result, the different displacement is equivalent to an overall loss of about 5 minutes between the width of the vortices in both sets of orbits. The implication here is that we can estimate the corresponding reduction of the vortex width across the wake between 2006 and 2013. Since the average VEX speed around Venus is nearly 7 km/s the spacecraft would reduce an equivalent ~2100 km travel distance difference in a 5 min travel time difference by moving through a vortex structure in the 2006 orbits with respect to those in the 2013 orbits. Thus, it is possible to argue that the vortex width decreases by nearly a ~ 2100 km distance across the wake between two different positions separated by a ~ 1 $R_V$ distance at ~1.5 $R_V$ and at ~2.5 $R_V$ along the X axis between both orbit sets.

Since the mass flux of planetary ions that move across a vortex is given by: $F = nUA$ where $A \sim r^2$ is the area of the vortex structure with r being the transverse distance to the vortex it is possible to require a constant value for the area integrated mass flux across the entire vortex as it decreases its size along the wake. By having a smaller distance r across the vortex the total mass flux will be more concentrated in the inner wake and hence larger values of the flow speed U will result (provided the density remains unchanged). For example, if we assume that in the $\Delta T \sim 12$ min time span for the orbit set where the vortex width is closer to Venus (right circle) so that $r_1 \sim 6000$ km (~1 $R_V$) at $X \sim -1.5\ R_V$, and that further downstream (by $X \sim -2.5\ R_V$) the width decreases to $r_2 \sim 3000$ km (so that $\Delta T \sim 6.5$ min as it would be suitable for the left circle), we can infer that $\Delta r \sim 3000$ km is the transverse distance decrease ($\Delta r \sim R_V/2$) between both orbit sets. With such a smaller cross-section size the area integrated mass flux indicated above is now compressed and thus in order to maintain its same value the flow speed should be larger by a factor of 4.

Such a speed increase is to be compared with the increase of the speed of the planetary ions that is estimated from the speed profiles in the right panels of **Figures 1** and **2** where it rises from ~10 km/s by the ~5000 km altitude to ~40 km/s by ~8000 km. The agreement between both variations supports the view that the enhanced speed values in the right panel of **Figures 1** and **2** are related to the VEX motion through the narrow section of a corkscrew flow as shown in **Figure 3**.

## 6. Conclusions

The peak mass flux values in the noon-midnight meridian in **Figure 4** occur by the 8000–10,000 km altitude range where large speed values of the planetary ions in **Figures 1** and **2** are measured. That correlation may be related to the fact that as the VEX spacecraft moves into the narrow region of the corkscrew flow (which is depicted in **Figure 3**) it will reach a region where the main bulk of the ionospheric plasma eroded by the solar wind is driven as it moves along the wake. The geometry of the corkscrew flow thus provides a suitable manner in which the eroded plasma that moves into the wake can be accumulated as it moves in that direction. As a result, high mass flux values of that flow should be encountered in the narrow section of the corkscrew flow geometry thus providing an account for the large values measured at high altitudes (by ~8000 km) shown in the right-side panel of **Figure 4**.

The high mass flux values identified at upper altitudes in a region way above the nightside ionosphere represent an implication obtained by forcing that plasma to move into the wake by decreasing its cross-section as in a fluid dynamic analog similar to that of **Figure 3** [8]. As a result, the geometry of the resulting corkscrew flow will ensure that such plasma will be confined within an ever-decreasing cross-section that, in turn, will accelerate the planetary ions that stream through that region.

It should also be noted that notable changes in the density and speed of the planetary ions cannot be clearly identified on the dawn-dusk meridian but mostly in the noon-midnight meridian. The implication here is that the latter represents a phenomenon that is more appropriate by the midnight meridian where the solar wind-driven planetary ions are compressed with the narrower cross-section of the region where they flow.

## Acknowledgements

We wish to thank Gilberto Casillas for the technical work provided. Financial support was available from the UNAM-IN108814-3 Project.

## Author details

Hector Pérez-de-Tejada[1]* and Rickard Lundin[2]

1 Institute of Geophysics, UNAM, México

2 Swedish Space Research Institute, Kiruna, Sweden

*Address all correspondence to: hectorperezdetejada@gmail.com

## References

[1] Lundin R et al. Ion flow and momentum transfer in the Venus plasma environment. Icarus. 2011;**215**:7

[2] Pérez-de-Tejada H et al. Plasma vortex in the Venus wake. Eos. 1982;**63**(18):368

[3] Pérez-de-Tejada H, Lundin R, Intriligator D. Plasma vortices in planetary wakes, chapter 13. In: Olmo G, editor. Open Questions in Cosmology. London, UK: IntechOpen; 2012

[4] Lundin R et al. A large scale vortex in the Venus plasma tail and its fluid dynamic interpretation. Geophysical Research Letters. 2013;**40**(7):273

[5] Pérez-de-Tejada H, Lundin R. Solar cycle variations in the position of vortex structures in the Venus wake, chapter 3. In: Bouvaquia, editor. Solar planets and Exoplanets. London, UK: Intech-Open; 2021. DOI: 10.5772/96710.2021

[6] Pérez-de-Tejada H, Lundin R. In: Barkitas I, editor. Vortex Dynamics in the Wake of Planetary Ionospheres: From Physical to Mathematical Aspects. London, UK: IntechOpen; 2022. DOI: 10.191252

[7] Perez-de-Tejada H, Lundin R. Particle Acceleration in the Corkscrew Flow within the Venus Plasma Wake. London, UK: IntechOpen; 2023

[8] Pérez-de-Tejada H. Distribution of plasma and magnetic fluxes in the Venus near wake. Journal of Geophysical Research. 1986;**91**:8039

[9] Pérez-de-Tejada H et al. Measurement of plasma channels in the Venus wake. Icarus. 2019;**321**:1026-1037

[10] Pérez-de-Tejada H et al. Plasma measurements of the PVO in the Venus ionosheath: Evidence for plasma heating near the ionopause. Journal of Geophysical Research. 1985;**90**(A2):1759-1764

[11] Romanov SA et al. Interaction of the solar wind with Venus. Cosmic Research. 1979;**16**:603

[12] Verigin M et al. Plasma near Venus from the Venera 9 and 10 wide angle analyzer data. Journal of Geophysical Research. 1978;**83**:3721

[13] Phillips J, McComas D. The magnetosheath and magnetotail of Venus. Space Sciences Rev. 1991;**55**:1

[14] Bridge A et al. Plasma and magnetic fields observed near Venus. Science. 1967;**158**:1669-1673

[15] Vörös Z et al. Intermittent turbulence, noisy fluctuations and wavy structures in the Venusian magnetosheath and wake. Journal of Geophysical Research. 2008;**113**:ED0B21. DOI: 1029/2008JE003159.200

[16] Delva M et al. Proton cyclotron waves in the solar wind at Venus. Journal of Geophysical Research. 2008;**113**:E00B06

[17] Dobe Z et al. Interaction of the solar wind with unmagnetized planets. Physical Rev. Lett. 1999;**83**(2):260-263

[18] Shapiro V, Shevchenko V. Astrophysical plasma turbulences. Astrophys. and Space Science Rev. 1968;**6**:427

[19] Szego K et al. Physical processes in the plasma mantle of Venus. Geophysical Research Letters. 1991;**18**(12):2305

[20] Luhmann J et al. A comparison of induced magnetotails of planetary

bodies; Venus, Mars, titan. Journal of Geophysical Research. 1991;**96**:1199

[21] Russell CT. Limits on the possible intrinsic magnetic field of Venus. Journal of Geophysical Research. 1980;**85**:8319

Chapter 3

# Magnetic Skyrmions and Quasi Particles: A Review on Principles and Applications

*Birhanu Abera Kolech*


## Abstract

Skyrmions are topologically nontrivial, magnetic quasiparticles that are characterized by a topological charge. The field of magnetic skyrmions has been actively investigated across a wide range of topics during the last two decades. We mainly reviewed and discussed magnetic skyrmions, and quasiparticles: reviews on principles and applications. We concentrated on theoretical discoveries and advances in magnetic skyrmions, topological effects, the skyrmion Hall effect, and the dynamics of skyrmions. The skyrmion Hall effect causes a transverse deflection of skyrmions when they are driven by currents, whereas the first additional contribution to the Hall effect of electrons in the presence of a topologically non-trivial spin texture may become beneficial for detecting skyrmions. This means that when a current is applied along the racetrack, skyrmions are pushed toward the edge, causing pinning or possibly the loss of data. This is one of the reasons why there is currently no prototype for a spintronic device based on skyrmions. The anti-ferromagnetic and ferrimagnetic skyrmions were then discussed in relation to spintronics. Finally, we reviewed several potential applications based on magnetic skyrmions, including skyrmion race track memory, a skyrmion logical device, a skyrmion magnonic crystal, and skyrmion-based radio frequency devices.


**Keywords:** spintronic device, topological Hall effect, skyrmion hall effect, quasi particles, bimeron, ferromagnetic, antiferromagnetic skyrmions

## 1. Introduction

Information technology has significantly increased in relevance for our daily lives during the past few decades. The demand for energy-efficient data manipulation and storage has recently increased due to the recent dominance of modern information technology applications like streaming services and cloud storage. Although Moore's law is difficult for present electronic solutions to follow [1], new spintronic ideas have been put out and might soon become important [2].

The racetrack memory is one of the most promising and anticipated data-storing systems. The bits, which are encoded by the presence or absence of the magnetic object, are written, erased, relocated, and read on a restricted track. This method was first proposed for using domain walls as carriers of information [3–5]. This almost one-dimensional configuration can be stacked, opening the door to naturally three-dimensional data storage with far higher bit densities. In terms of lower energy consumption and quicker access times, this non-volatile idea outperforms conventional random access memory and hard disk drives [4].

This review focuses on non-collinear spin textures because topological matter, a promising study area outside of spintronics, is a focus of this work. The magnetic skyrmion [6] is the most notable example. A decade ago, this whirl-like nano-object was discovered [7]. Due to its topological protection, which offers it a tremendous amount of stability even at small scales, it has the potential to be an information carrier in the next generation of data storage devices, such as racetrack nano-devices [8–10]. In addition to their high stability, the topological characteristics of skyrmions also lead to emergent electrodynamics, such as the topological Hall Effect and the skyrmion Hall Effect [11, 12]. The skyrmion Hall effect causes a transverse deflection of skyrmions when they are driven by currents, whereas the first an additional contribution to the Hall effect of electrons in the presence of a topologically non-trivial spin texture may become beneficial for detecting skyrmions [13, 14]. This means that when a current is applied along the racetrack, skyrmions are pushed toward the edge, causing pinning or possibly the loss of data. This is one of the reasons why there is currently no prototype for a spintronic device based on skyrmions.

While work to expand the use of magnetic skyrmions in spintronic devices will continue, various alternative nano-objects have been predicted and seen over the past six years. Research in this area will be further accelerated in the near future, as some of them hold even greater advantages over traditional skyrmions. We outline and discuss these additional magnetic quasiparticles in this review. We categorize the objects, describe how to stabilize them, and contrast their emergent electrodynamics with that of typical skyrmions.

## 2. Magnetic skyrmions

Tony Skyrme first proposed the existence of skyrmions in the context of particle physics in the 1960s. He demonstrated the existence of particle-like solutions with the characteristics of baryons and presented a field-theoretical explanation of interacting pions [15, 16]. Later, it was demonstrated that whereas pions themselves are bosonic, these solutions exhibit Fermi properties [17, 18]. The three-dimensional equivalents of what came to be known as "skyrmions" are the solitons, which are explained by a non-linear sigma model.

Skyrmions have recently been discovered in a number of physics domains, including string theory [19], liquid crystals [20], Bose-Einstein condensates [14], quantum Hall systems [13], and magnetism [7].

A skyrmion can be thought of in this context as a two-dimensional object that is trivially continued along the third dimension (**Figure 1a** and **c**). According to reciprocal-space measurements [7] and Lorentz transmission electron microscopy [23], these skyrmion tubes or strings were first discovered in MnSi in 2009 [23]. These objects' magnetic textures confirmed the predictions made twenty years earlier [6]. Magnetic skyrmions in a ferromagnetic medium produce a non-trivial real-space topology in its core due to a constantly changing magnetization density that is oriented in opposition to the surroundings. These substances can exist as discrete particles or periodic lattices, as in the publications mentioned above [23, 24].

As these topics are necessary to comprehend the physics of the alternative magnetic quasiparticles, the treatment of conventional skyrmions is restricted to their

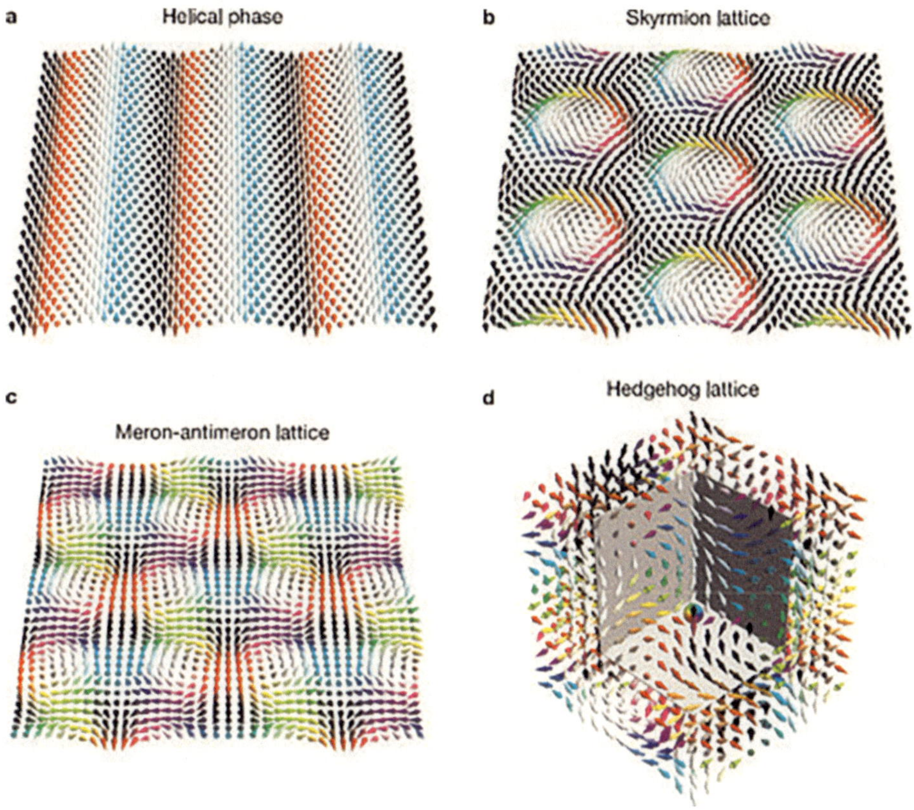


**Figure 1.**
*Changing textures periodically. (a) A single Q vector designates the helical phase. (b) As seen in [21], a skyrmion lattice or skyrmion crystal is formed by superimposing three helices with Q vectors angled at a 120° angle. According to MnSi, a B20 material, the skyrmions (vorticity m = 1) are of the Bloch type (helicity $\gamma = \frac{\pi}{2}$) and form a hexagonal superlattice. Antimerons (vorticity m = −1) with a negative net magnetization (black) and merons (vorticity m = 1) with a positive net magnetization try to compensate the meron-antimeron lattice in (c) (white). The lattice has a positive net topological charge because both objects have a topological charge of NSk = +1/2. (d) The hedgehog lattice, also known as the Bloch anti-Bloch crystal, is created by superimposing three helices in three dimensions. The magnetization density in a continuous description has singularities. These Bloch or anti-Bloch points do not have a definite magnetization. A colored sphere, the Bloch point, is used to draw attention to one of these points. Source: Göbel B, Mertig I, Tretiakov OA. Beyond skyrmions: Review and perspectives of alternative magnetic quasiparticles [Internet]. Vol. 895, Physics Reports. Elsevier BV; 2021. p. 1–28. Available from: http://dx.doi.org/10.1016/j.physrep.2020.10.001 [22]. © 2021 The Author(s). License: CC-BY 4.0. Used with permission.*

geometrical characterization, stabilizing processes, and emerging electrodynamics. We cite one of the many review articles [25, 26] for a discussion of conventional skyrmions that goes beyond these issues.

## 2.1 History of Skyrmions in magnetism

For a very long time, skyrmions in magnetism have been explored. The most well-known magnetic Skyrmions are magnetic bubbles. From the 1960s through the 1980s, there was a lot of research done on these circular domains in an out-of-plane magnetized medium, partly because of the potential for use in solid state storage devices [27], which eventually led to commercial devices [28]. Industrial interest in bubble media was eventually lost due to the increasing efficiency of rotating hard disks in the 1980s, which was aided by the discovery of the gigantic magneto-resistance in 1988 [29, 30] and the development of flash memories [31]. However, research on current-induced domain wall motion [32] and racetrack memory devices [4] shows that the old idea of a store medium that is based on domains moving in a solid state device is once again receiving a great deal of attention. Skyrmions are promising candidates for the realization of such a device because of their remarkable mobility at extremely low currents, and it would be advantageous to repurpose the rich expertise of bubble-based technology. This thesis study, in which we examine the dynamics of magnetic bubbles at short (ns) time scales for the first time, could help to establish a connection between established bubble physics and cutting-edge work on domain wall physics for use in practical applications.

Kooy and Enz's work, which discovered an accurate theoretical model for the energetics and evolution of stripe domains and bubbles under the application of an external magnetic field [33], served as the catalyst for the initial research on bubble domains. The creation and manipulation of controlled bubbles began with this static model. The experimental development of operational bubble-based devices was afterward largely led by Andrew Bobeck and his Bell Laboratories team [27].

On the theoretical front, two turning points in our knowledge of bubble-domain dynamics must be emphasized. First, the one-dimensional model, developed to describe the motion of bubbles and which we will examine in more depth below, is quite successful in modeling the straight motion of magnetic domain walls, despite certain shoddy assumptions. The one-dimensional model is currently regarded as the accepted explanation for domain wall motion. The addition of the gyrocoupling vector to the equation describing the motion of magnetic bubbles was the second significant theory [34, 35]. The observation of the so-called skew deflections of bubbles [36] served as the impetus for Thiele's studies. It was discovered that the steady state velocity of a bubble has a sizable component perpendicular to the gradient rather than generally following the field gradient. Every bubble has a variable deflection angle, and it was even unpredictable whether the bubble would be redirected from the field gradient direction to the right or to the left.

## 2.2 Topology and characterization

A stereographic projection helps to explain the topological nature of a skyrmion. By rearranging the magnetic moments of a three-dimensional hedgehog, where all moments on a sphere point in the radial direction; a two-dimensional skyrmion can be created. Without altering the direction of the moments, this sphere gets split open at the bottom and flattened into a disk. The outcome is a two-dimensional magnetic entity that is topologically non-trivial (**Figure 2**).

The very first three products are various kinds of skyrmions, meaning they have different helices and vorticities: (a) an antiskyrmion with a vorticity of m = 1 and a topological charge of $N_{Sk}$ = 1; (b) a skyrmion with an intermediate helicity of $\gamma = \frac{\pi}{4}$ among both Bloch and Néel type skyrmions and a topological charge of $N_{Sk}$ = 1; and (c) a (d) Illustrates a magnetic bimeron made of two merons. Another interpretation is that it is a skyrmionic excitation in an in-plane magnetic medium, where $N_{Sk}$ = 1. In the center row, two skyrmions are combined: (e) a biskyrmion with $N_{Sk}$ = 2, (f) a skyrmionium with $N_{Sk}$ = 0, and (g, h) ferrimagnetic and synthetic antiferromagnetic skyrmions, for which the topological charges of the two subskyrmions balance one another. The skyrmion tubes (possibly with varying helicity along the tube), the chiral bobber as a discontinued skyrmion tube, a pair of Bloch and anti-Bloch points serving as the building blocks of a three-dimensional crystal (hedgehog lattice), and the

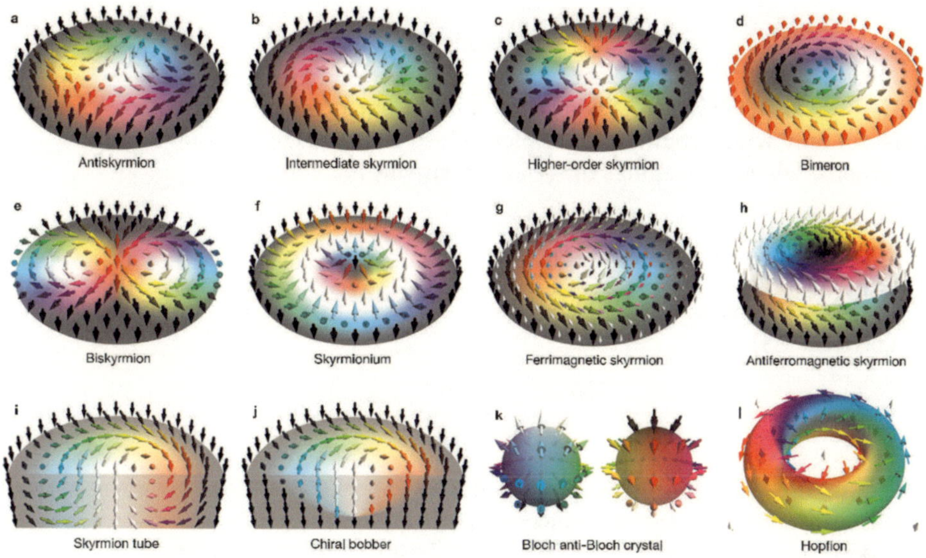


**Figure 2.**
*Overview of the discussed topologically non-trivial spin textures. Source: Göbel B, Mertig I, Tretiakov OA. Beyond skyrmions: Review and perspectives of alternative magnetic quasiparticles [Internet]. Vol. 895, Physics Reports. Elsevier BV; 2021. p. 1–28. Available from: http://dx.doi.org/10.1016/j.physrep.2020.10.001 [22]. © 2021 The Author(s). License: CC-BY 4.0. Used with permission.*

hopfion are all shown as extensions of skyrmions in the bottom row. The magnetic moment orientation is depicted by colored arrows. The positive and negative out-of-plane orientations are represented by white and black, respectively [25].

A skyrmion with a magnetization density of $m(\mathbf{r})$ in a continuous picture cannot shift to a ferromagnetic state without discontinuous changes in the density. This is an example of the topology of real space that is not trivial as measured by the topological charge:

$$N_{SK} = \int n_{SK}(\boldsymbol{r})d^2r \tag{1}$$

which is as an integral over the topological charge density

$$n_{SK} - \frac{1}{4\pi}\boldsymbol{m}(\boldsymbol{r}).\left[\partial\frac{\boldsymbol{m}(\boldsymbol{r})}{\partial x}x\partial\frac{\boldsymbol{m}(\boldsymbol{r})}{\partial y}\right] \tag{2}$$

The following transformation makes it easier to infer a skyrmion's topological charge from its appearance. One expresses the magnetization density in spherical coordinates with the azimuthal angle θ and the polar angle Φ and expresses the position vector in polar coordinates $\mathbf{r} = r(\cos\varphi, \sin\varphi)$. Exploiting the radial symmetry of the out-of-plane magnetization density $\theta = \theta(r)$, the topological charge reads [25]

$$N_{SK} = \int_0^\infty dr \int_0^{2\pi} d\varnothing \frac{\partial\varnothing(\varnothing)}{\partial\varnothing}\frac{\partial\theta(r)}{\partial r} sin\theta(r) = -\frac{1}{2}cos\theta(r)\Big|_{r=0}^{\infty} \cdot \frac{1}{2\pi}\varnothing(\varnothing)\Big|_{\varnothing=0}^{2\pi} \tag{3}$$

The out-of-plane magnetization of a skyrmion is reversed comparing its center with its confinement. This is quantified by the first factor, the polarity

$$p = -\frac{1}{2}cos\theta(r)|\begin{matrix}\infty \\ r=0\end{matrix} = \pm 1 \tag{4}$$

The sign is dependent on the skyrmion host's out-of-plane magnetism. The polar angle can only wrap around in multiples of 2 since the magnetization density is continuous, which determines the second element, the vorticity.

$$m = \frac{1}{2\pi}|\begin{matrix}2\pi \\ \varnothing = 0\end{matrix} = 0, \pm 1, \pm 2, \ldots.. \tag{5}$$

The two-dimensional integral has been simplified to a product of the polarity and the vorticity [33]

$$N_{SK} = m.p = \pm 1, \pm 2, \ldots. \tag{6}$$

$$\Phi = m\varphi + \gamma \tag{7}$$

with an offset γ. This quantity is called helicity.

We have developed the three characteristic quantities for the various types of skyrmions, which are commonly stated as polarity, vorticity, and helicity.

$$N_{SK}(r) = \begin{pmatrix} \left(\frac{x\ cos\,\gamma}{r} - m\frac{y}{r}\,sin\,\gamma\right)\sin\left(\frac{\pi}{r_o}r\right) \\ \frac{x\ sinr}{r} + m\frac{y}{r}\ cos\,\gamma)\sin\left(\frac{\pi}{r_o}r\right) \\ pcos\left(\frac{\pi}{r_o}r\right) \end{pmatrix} \tag{8}$$

For 0 < r < $r_o$. Note, that the out-of-plane magnetization profile is simplified as a cosine function (radius r0) and that the exact profile depends on the interaction parameters, the sample geometry, defects, and the presence of other quasiparticles.

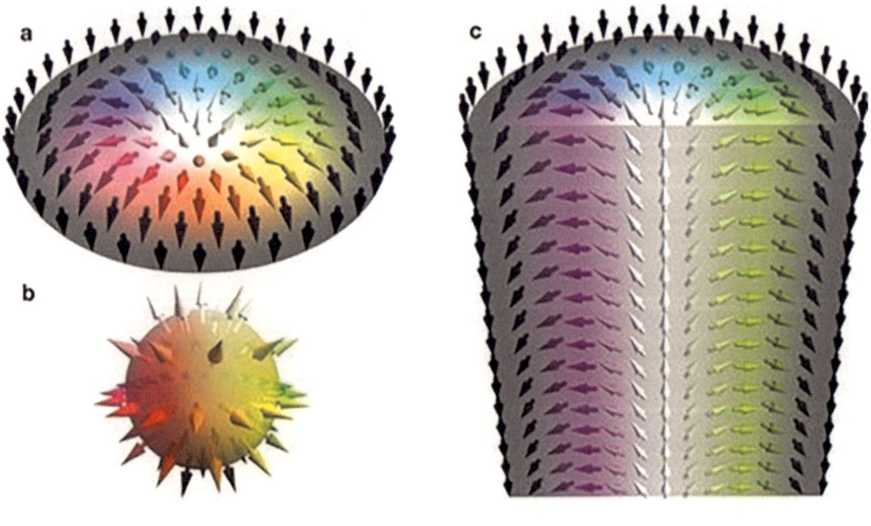


**Figure 3.**
*Magnetic skyrmions are fundamentally non-trivial excitations in ferromagnets. (a) A two-dimensional magnetic symmetry; (b) a three-dimensional Bloch point or hedgehog that is stereographically projected to form (c) The skyrmion frequently extends trivially in three dimensions as a skyrmion tube. Source: Göbel B, Mertig I, Tretiakov OA. Beyond skyrmions: Review and perspectives of alternative magnetic quasiparticles [Internet]. Vol. 895, Physics Reports. Elsevier BV; 2021. p. 1–28. Available from: http://dx.doi.org/10.1016/j.physrep.2020.10.001 [22]. © 2021 The Author(s). License: CC-BY 4.0. Used with permission.*

As an example, the skyrmion in **Figure 3a** has a positive polarity p = +1 and vorticity m = +1 leading to a topological charge of NSk = +1. Since the in-plane component of the magnetization is always pointing along the radial direction, the helicity, in this case, is $\gamma = 0$. This type of skyrmion is called Néel skyrmion and is typically observed at interfaces [37]. Different from this type of skyrmions are the skyrmions in MnSi (e. g., from the initial observation [7]). They are called Bloch skyrmions. There, the in-plane components of the magnetization density are oriented perpendicularly with respect to the position vector. This toroidal configuration is characterized by a helicity of $\gamma = \pm\pi/2$. In contrast to the polarity and the vorticity, the helicity is a continuous parameter allowing for skyrmions as intermediate states between Bloch and Néel skyrmions, as shown in **Figure 1b**. Furthermore, the vorticity can in principle take any integer value constituting, for example, antiskyrmions for m = −1 (**Figure 1a**) or higher-order (anti)skyrmions for $|m| > 1$ (**Figure 1c**). Out of this manifold, Bloch [7], Néel skyrmions [37], and skyrmions with an intermediate helicity [38], as well as antiskyrmions [39] have been observed experimentally. Higher-order skyrmions [21, 40] have been predicted.

## 2.3 Skyrmions in more materials

### *2.3.1 Bulk crystals*

A flurry of fresh trials on various materials was conducted after this initial observation. **Table 1** provides a summary.

The list includes everything from metals [7, 41–43, 52] to doped semiconductors [50, 51] to insulators [55–60]. FeGe has the highest critical temperature on the list because a skyrmion lattice can nearly always be seen at room temperature. The magneto-electric properties of the insulator $Cu_2OSeO_3$, which allow it to be controlled by both magnetic and electric fields [57, 72–74], are one of its specialties. These discoveries were accompanied by magnetic [75–78] and ultrasonic [79] resonance studies, which show good agreement between theory and experiment.

A skyrmion lattice was also seen in 2015 by Tokunaga et al. [63] in -Mn-type Co-Zn Mn alloys. These alloys exhibit broken inversion symmetry in the bulk, however, they have space groups $P4_132$ or $P4_332$ instead of the materials' $P2_13$ space group. It is interesting to note that the altered composition causes both a bigger length scale for the magnetic structures and a higher temperature for the observation of the skyrmion lattice. At room temperature [62] and even higher [63], skyrmions can be seen depending on the composition ($Co_xZn_yMn_z$ with x + y + z = 20).

However, the thermodynamically stable skyrmion lattice can be found in a very tiny pocket of the (universal) B-T-phase diagram, which is a feature shared by all bulk magnets. This finding is consistent with theoretical hypotheses supported by both analytical [7, 80] and numerical study [81] research (Monte Carlo). One method for expanding the pocket is to take advantage of the skyrmion's (topological) stability. Bloch points, also known as emergent monopoles, are single and thus energy-dense spin configurations that must destroy one skyrmion string or join two skyrmions together [82–85].

## 2.4 The interactions of skyrmions

Skyrmions interact with edges, flaws, magnetic textures, and especially other skyrmions while they are inside the nanostructure. According to simulations, skyrmions are attracted to both edges and other skyrmions [86–88].

| Material | Sample | conduction | $\Delta T_{sky}/K$ | λH/nm | Type | Refs. |
|---|---|---|---|---|---|---|
| MnSi | Bulk | Metal | 28–29.5 | 18 | Bloch | [7, 41–43] |
| MnSi(press) | Bulk | Metal | 5–29 | 18 | Bloch | [44–47] |
| MnSi | Film(~ 50 nm) | Metal | <5–23 | 18 | Bloch | [48, 49] |
| $Fe_{1-x}Co_xSi$ | Bulk | Semi -metal | 25–30 | 37 | Bloch | [50, 51] |
| $Fe_{0.5}Co_{0.5}Si$ | Film(~ 20 nm) | Semi- metal | 5–40 | 90 | Bloch | [23] |
| FeGe | Bulk | Metal | 273–278 | 70 | Bloch | [52] |
| FeGe | Film(~ 75 nm) | Metal | 250–270 | 70 | Bloch | [52, 53] |
| FeGe | Film(~ 15 nm) | Metal | 60–280 | 70 | Bloch | [54] |
| $Cu_2OSeO_3$ | Bulk | Insulator | 56–58 | 60 | Bloch | [55–60] |
| $Cu_2OSeO_3$ | Film(~ 100 nm) | Insulator | <5–57 | 50 | Bloch | [56, 61] |
| $Co_8Zn_8Mn_4$ | Bulk | Metal | 284–300 | 125 | Bloch | [62] |
| $Co_8Zn_9Mn_3$ | Bulk | Metal | 311–320 | >125 | Bloch | [63] |
| $Co_8Zn_9Mn_3$ | Film(~ 150 nm) | Metal | 300–320 | >125 | Bloch | [63] |
| $GaV_4S_8$ | Bulk | Semi- metal | 9–13 | 17.7 | Nêel | [64] |
| Fe/Ir/(111) | Monolayer | — | ~ 11 | ~ 1 | Nêel | [38, 65, 66] |
| PdFe/Ir/(111) | Bilayer | — | ~ 4.2 | ~ 1 | Nêel | [65–70] |
| $(Ir/Co/Pt)_{10}$ | Multilayer | Metal | *RT* | 30–90 | Nêel | [71] |
| Pt/CoFeB/MgO | Multilayer | Metal | RT | 480 | Nêel | [71] |

*The second column in the table includes information on bulk materials, thin films, and layer materials. The third column contains information about the conductivity, if it is known. The temperature range $T_{sky}$ at which skyrmions were spotted, the helical/stripe phase's wavelength λH, and the skyrmion's texture are all listed in the ensuing columns.*

**Table 1.**
*Materials chosen that are known to contain skyrmion lattices or individual skyrmions.*

This is true only if the trade relationship is not stifled, though [21, 89]. Additionally, if the backdrop is not out-of-plane polarized but rather polarized in another direction [90, 91] and conical [92, 93], the interaction of skyrmions with both other skyrmions and edges becomes appealing.

The idea to also include the racetrack's surface was just recently put out, as in the suggestion in Ref. [94], where it was demonstrated that a scratch in the surface can draw skyrmions and serve as a track without physical edges.

## 2.5 Pushing skyrmions

The concept of using spin-polarized currents to move the skyrmions along the track has been addressed [95–98]. Keep in mind that the mass of the skyrmions has no bearing in these circumstances. It should be emphasized, however, that Dzyaloshinsky-Moriya interaction-stabilized skyrmions are more rigid and have a low mass [97], whereas huge skyrmions, which typically require strong dipolar contacts, can readily deform and have a non-negligible mass [99, 100].

Due to their interaction with magnons, skyrmions may also be moved in this manner. Early simulations revealed that skyrmions in a temperature gradient gravitate toward the source of heat [101], which may be regarded as a source of magnons.

More in-depth investigations revealed [102–106] that the source of magnons does, in fact, draw skyrmions. This theory can be applied to control the skyrmions in nanostructures [107]. For more information, however, it should be noted that the interaction of magnons with edges and other textures is non-reciprocal in nature [108, 109]. A tilted background field combined with an oscillating superposition, which breaks enough symmetry once again to occupy the translational mode of a skyrmion, makes up a more macroscopic but related mechanism [110].

The stability of the skyrmions' trails has also been researched. Equidistant skyrmions should cover the same distances in the same amount of time if a storage device uses them. With the exception of interactions between skyrmions, which were previously thought to be completely pure, this can be assumed to be true. Studies on interactions between skyrmions and random defects such as vacancies or holes, as well as localized changes in magnetic properties, revealed that interactions between skyrmions and defects can exhibit behavior other than repulsive behavior. As calculated in effective particle models [92] and simulations of tracks made up of patches with changing anisotropy, the motion of a skyrmion can appear unexpected when combined with the gyroscopically dominated dynamics, as predicted from tests [111] (**Figure 4**).

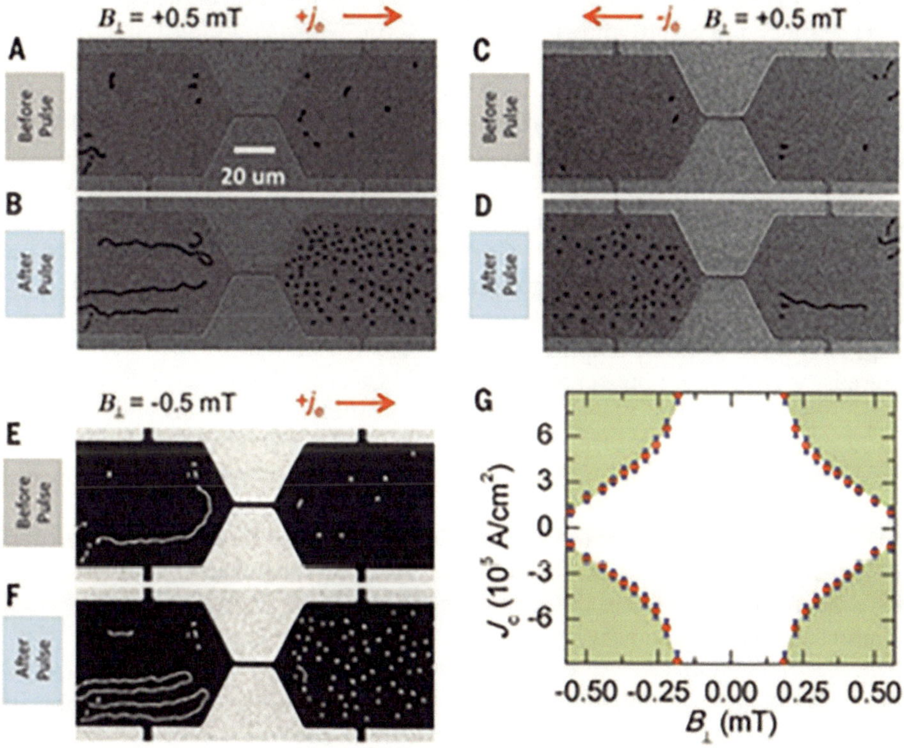


**Figure 4.**
*Observational research as described by Jiang et al. [112]. We scanned and nanostructured a trilayer made up of Ta (5 nm)/Co20Fe60B20(CoFeB) (1.1 nm)/TaOx (3 nm). In panels (E) and (F), the trilayer-covered portions stand out as distinct black zones. These are skyrmions or extended skyrmions, which are the smaller structures in this region (bimeronsThe trilayer is exposed to a magnetic field B that is perpendicular to it. Note that the magnetic field is pointing in the opposite direction in (E) and (F), and as a result, the contrast is reversed. The initial states before any current is applied are shown in (A), (C), and (E). The magnetization is shown in (B), (D), and (F) following the application of a current pulse in the direction denoted by the red arrows. As a result, skyrmions gather in the direction of the current. The skyrmion proliferation is seen above the current densities Jc and fields B, as indicated by the green patches in the phase diagram (G) [112].*

The thermal diffusion of skyrmions may potentially result in other issues [97]. Here, it is suggested that the racetrack be divided into parking lots utilizing a variety of techniques, such as voltage gating [113]. For skyrmion systems, antiferromagnetic coupling of layers has been proposed to remove the complex gyroscopic motion [114], which is well known for its beneficial effects on acceleration in conventional domain wall racetracks. Other ideas include antiferromagnetic rather than ferromagnetic nearest neighbor exchange interactions [115], which already solves the issue of stray fields.

## 3. Topological Hall effect, Skyrmion Hall effect, and Skyrmion dynamics

The internal structure of magnetic materials may have some peculiar features that have been the subject of decades of research. One illustration is the Hall Effect. Due to this finding, applying a magnetic field to a conductor that is carrying current causes a voltage drop across the conductor that is perpendicular to the current and transverse to both the applied magnetic field and the current [116]. When a ferromagnet is taken into account, the related influence on the Hall resistivity has two terms. The charged particles deflecting as they move with a velocity perpendicular to the applied field cause the first component, known as the Hall term, which is proportional to the applied field. The anomalous or spontaneous Hall term, the second term, is inversely proportional to magnetization. Smit and Volger [117] concluded that this term is likely caused by the ferromagnets internal magnetic field, which is caused by dipoles. This term was then linked to the spin-orbit effect by Smit [118]. It was also known that electrons in a magnetic field that were hopping between atoms might develop a phase factor. The Berry phase is the common name for this variable, which is dependent on the magnetic vector potential [119]. This phase component in a chiral magnet would depend on the effective field owing to the spin chirality, which would depend on the topological and geometrical characteristics of the lattice, as realized by Taguchi et al. [120]. This phase could be empirically seen as changes in the conductivity of chiral materials, as realized by Binz et al. [121]. In other words, the system's chiral structure, or topology, would cause a Hall effect. This is significant because, as will be discussed shortly, similar topological effects have been found in substances like MnSi that may host skyrmions.

The topological Hall effects in MnSi were seen by Neubaeur et al. [41] and Lee et al. [44]. Under a variety of pressures, Lee et al. observed a stepwise field profile in the Hall conductivity, while Neubaeur et al. [41] observed a Hall effect in the T-B areas, which corresponded to the A-phase of MnSi, or the skyrmion lattice phase. The variations in Hall conductivity, however, were not like the typical anomalous Hall effect. Neubaeur et al. came to the conclusion that the reason for the unusual conductivity behavior was the coupling of the applied current's spins to the chiral skyrmion lattice. The finding was that the effective internal magnetic field created by the skyrmion lattice caused the conduction electrons that make up the current to acquire a Berry phase. The skyrmion lattice's effective field in standardized unit vectors is [41]:

$$B^{\mu} = \frac{1}{8\pi}\frac{h}{e}\epsilon_{\mu\upsilon\lambda}\hat{n}.(\partial_{\upsilon}\hat{n}x\partial_{\lambda}\hat{n}) \tag{9}$$

i.e., a magnetic field that is proportional to the topological charge. Now $\epsilon_{\mu\upsilon\lambda}$ is the antisymmetric tensor. The topological Hall effect for skyrmions causes a voltage drop in the direction opposite to the applied current and the magnetic field, which is itself perpendicular to the current, much like the normal Hall effect does [41].

Zang et al. [122] explored the associated Skyrmion Hall Effect analytically and quantitatively. They used the conventional exchange, Zeeman, and DMI energies to investigate the collective dynamics of a skyrmion lattice in a thin film. In the end, Zang et al. discovered that the introduction of an electric current could move the skyrmion lattice as a whole, and that the skyrmions would move with a trajectory at an angle to the current. The Skyrmion-Hall effect is the name given to these phenomena. By examining deformations of the skyrmion lattice in terms of a stiff approximation of the spin vectors, such that n (r, t) = n(r u(r, t)), Zang et al. considered this process analytically. This holds true for an elastic deformation that slowly varies with regard to the skyrmion lattice scale.

As stated by Zang et al. [122], the skyrmion lattice as a whole, which in the rigid approximation will have a velocity$\dot{\boldsymbol{u}} = \boldsymbol{v}$ , is driven by an applied current. The subscript denotes that this is the component of the velocity in the direction of the current. The moving skyrmion lattice will then induce an internal electric field,$\boldsymbol{E} = \frac{1}{c}\boldsymbol{v}\|\boldsymbol{x}\boldsymbol{B}$ due to its internal structure, which has the internal magnetic field **B**. Hence, Zang et al. noted that applying a current to the skyrmion lattice will generate an electric current transverse to the direction of $\boldsymbol{v}\|$, and this current is determined by the internal magnetic field of the skyrmion, which in turn is determined by its topology. A voltage drop will result in a direction that is opposite to the direction of the applied current. This can be understood as the topological Hall effect for the moving skyrmion lattice, or the previously described skyrmion Hall effect (SHE), which causes the skyrmion lattice to move at an angle to the initially applied current, according to Zang et al. [122]. Moreover, Zang et al. [122] showed that the transverse component of the skyrmion lattice velocity, $\mathbf{v}\perp$ is due to dissipation as a result of the skyrmion lattice's internal field. Specifically, Zang et al. showed that the dissipation arises from the coupling of the conduction electrons in the applied current with the local magnetic moments of the film. One can define the corresponding skyrmion Hall angle as $\theta = \frac{v}{v}$, where $\mathbf{v}\perp$ is proportional to the topological charge. The internal magnetic fields of the skyrmion are ultimately responsible for the topological and/or skyrmion Hall effects, which result in step-like behavior in the Hall conductivity or skyrmion trajectories that are perpendicular to the applied current, respectively. Be aware that Jiang et al. [112] directly observed the spin Hall effect in an experiment a few years later.

Skyrmions can be driven by very low current densities, several orders of magnitude smaller than domain walls, which is encouraging for potential applications even though the spin Hall effect is not ideal for applications involving skyrmions because the skyrmion will not be driven precisely along the current direction. For instance, Yu et al. [123] demonstrated this in the material FeGe at a temperature close to ambient. Yu et al. used Lorentz transmission electron microscopy to examine the translational and rotational motion of the skyrmion lattice, with the spin-transfer torque serving as the primary driving force. This procedure entails the transfer of spin via a polarized current, causing the skyrmion to experience a torque and move along as a result. According to Yu et al., the depinning of domain walls and the depinning of skyrmions from defects both require a tiny threshold current, but the threshold current required to depin skyrmions is substantially less. As was previously mentioned, the Skyrmion Hall effect will result in some transverse velocity.

Iwasaki et al.'s [95] investigation of the spin transfer torque's role in driving a skyrmion With the exception of the Gilbert factor being replaced by the time derivative of the magnetization, they numerically solved the Landau-Lifshitz-Gilbert equation, which is equivalent to the Landau-Lifshitz equation describing the dynamics of magnetization [124]. Iwasaki et al. showed numerically that there is a linear current-velocity relation for the spin transfer torque-driven skyrmion for the component of the velocity along the direction of the applied current for the system with the exchange, Zeeman, Dzyaloshinsky-Moriya interaction, and random anisotropy.

This velocity component is also independent of the Gilbert damping, impurities, and non-adiabatic effect due to the interaction of the spins of the magnetic moment with the spin-polarized current. This supports the conclusions reached by Zang et al. [122]. Iwasaki et al. hypothesized that this was due to the skyrmion's ability to alter its lattice structure or alter its individual shape in order to avoid being trapped by contaminants. However, because skyrmions rotate around impurity centers, impurities do affect the transverse component of velocity and, consequently, the skyrmion Hall angle.

In order to demonstrate that temperature gradients can also be used to drive skyrmions, Kong et al. [101] examined the dynamics of skyrmions in a thin layer. The Landau- Lifshitz-Gilbert equation with the exchange, Zeeman, and Dzyaloshinsky-Moriya interaction was numerically solved and thermal fluctuations were included as a random field to analyze this. The random fields caused the skyrmions to move. The velocity was random at any given time, according to Kong et al. The drift velocity, however, was not zero and was going in the opposite direction of the temperature gradient. Additionally, the velocity's longitudinal component correlated with the temperature gradient [101].

### 3.1 Antiferromagnetic skyrmions

In antiferromagnets [125, 126] and artificial antiferromagnetic bilayers [114], the dynamic characteristics and stability of single antiferromagnetic skyrmions were first predicted. They were soon expanded to include two-sub lattice antiferromagnetic skyrmion crystals [127]. Antiferromagnetic skyrmions can be thought of as the union of two skyrmions with mutually reversed spins, just as with skyrmioniums. As a result, they are distinguished by a vanishing topological charge. The subskyrmions in this instance, however, are not spatially separated but rather entangled. These results in the local disappearance of the magnetization density and the magnetization can be replaced by the Néel order parameter, which is the primary order parameter for antiferromagnets. Using this parameter to determine the topological charge yields a result of$\pm 1$. Thus, from the perspective of topology, antiferromagnetic skyrmions are still skyrmions, but they behave differently from ferromagnetic skyrmions. The Thiele equation, albeit with the Néel order parameter, can also explain these antiferromagnetic dynamics [128].

An antiferromagnetic skyrmion moves without the skyrmion Hall effect due to the compensated topological charge for magnetization, as proposed for skyrmionium [114, 125]. The two subsystems, on the other hand, are considerably more strongly connected and prevent a deformation brought on by a pairwise opposing transverse motion, as was the case for the skyrmionium. The antiferromagnetic skyrmions can be pushed by currents significantly more quickly than normal skyrmions, which is characteristic of antiferromagnetic spin textures. Simulated speeds in the kph range have been recorded [129].

Antiferromagnetic skyrmions are therefore the best information carriers for data storage systems. Additionally, it was theoretically demonstrated that, in contrast to their ferromagnetic counterparts, antiferromagnetic skyrmions have a high diffusion constant [125] in systems with low damping, suggesting a potential for driving them using temperature gradients. They also do not show stray fields, which would enable a denser stacking of almost one-dimensional racetracks when creating a three-dimensional storage device.

The stabilization of antiferromagnetic skyrmions is not difficult given the necessary Dzyaloshinsky-Moriya interaction [125, 130]. A skyrmion with mutually reversed spins—that is, one with opposite polarity and a helicity difference of—is also energetically stable when one takes into account the kind of skyrmion that the Dzyaloshinsky-Moriya interaction in a system prefers energetically (given by the symmetry). The antiparallel alignment of the respective magnetic moments also requires rather significant antiferromagnetic coupling between the two subskyrmions. In fact, the bilayer-type antiferromagnetic skyrmions have just lately been seen in room-temperature synthetic antiferromagnets [131, 132]. In Ref. [131], magnetic force microscopy was used to identify the tiny stray fields that resulted from the bilayer arrangement. The authors of Ref. [132] describe how to create synthetic antiferromagnets with a variable net moment. They prepared a system with a tiny net moment in addition to a fully corrected system so they could undertake magneto-optical Kerr effect experiments.

A regulated generation procedure is required for antiferromagnetic skyrmion-based logic [130] or racetrack [125] applications. For instance, conventional skyrmions have been produced via directed, deterministic methods like spin torques or magnetic fields [25]. However, these methods are challenging to apply to antiferromagnetic skyrmions because all vectorial quantities would either have to act on one of the two subskyrmions alone, causing the other to generate automatically due to the strong antiferromagnetic coupling, or they would have to act on both subskyrmions with the opposite sign. Both strategies are hardly practical since, for instance, magnetic fields cannot alter their sign on the lattice constant-length scale. In this case, stochastic processes seem to be more favorable, such as the creation of nano-objects at flaws or from confinement (as demonstrated for ordinary skyrmions [67, 133]). When an antiferromagnetic skyrmion crystal needs to be stabilized, the stability of antiferromagnetic skyrmions becomes considerably more difficult. Using skyrmion crystals as an example, a stabilizing magnetic field is necessary. For each of the two subsystems, it must be aligned along z. A hypothetical antiferromagnetic skyrmion host might be grown on top of a collinear anti-ferromagnet with the same crystal structure at the interface to get around this issue. The exchange interaction at the contact imitates a staggered magnetic field as a result [127].

Another problem is detecting antiferromagnetic skyrmions in a single layer (rather than in a synthetic anti-ferromagnet bilayer). Global and local compensation is made for both the magnetization and topological charge density of the magnetization (middle and bottom panels). For real-space techniques like magnetic force microscopy or Lorenz transmission electron microscopy, these antiferromagnetic skyrmions would consequently appear to be undetectable. Additionally, there are no aberrant or topological Hall signatures. Fortunately, a different characteristic, the topological spin Hall effect, has been proposed [133–137]. The resulting signal is an analog of the traditional spin Hall effect, but it comes from the spin texture's non-collinearity.

It is easiest to understand the topological spin Hall effect if one makes the assumption that there are two electronically uncoupled sub skyrmions. Because of the opposing spin orientation, the emergent fields of the two subskyrmions are oriented in opposition. The electrons are transversely deflected in opposing directions as a result. Due to the opposite spin alignment, the two species of electrons can be thought of as having "spin up" and "spin down" states depending on how well their spins line up with their respective textures. However, the foregoing concept of an emergent magnetic field becomes problematic because the sublattices are truly connected. Additionally, a non-Abelian formulation must be taken into consideration to account for the sublattice-degenerate bands in order to calculate the spin Hall conductivity from the reciprocal space properties. Similarly, the spin-polarization of an electron moving through an antiferromagnetic skyrmion is no longer perfectly aligned with the texture. The conduction electrons' orbital motion becomes significant and spin-dependent [138, 139]. Nevertheless, a topological spin Hall effect emerges.

In conclusion, there is a reason for optimism regarding the use of antiferromagnetic skyrmions in spintronic devices in the future. Recently, the current-driven motion of synthetic antiferromagnetic skyrmions has been accomplished [132]. Synthetic antiferromagnetic skyrmions have been seen. Skyrmions with a single layer of anti-ferromagnetism have also been suggested. Despite fully compensated magnetizations, stray fields, and topological charge densities of the magnetization, the topological spin Hall effect may be crucial for viewing these things. The skyrmions might also be moved by heat gradients or electrically generated anisotropy gradients, even in antiferromagnetic insulators, as was predicted.

As a last point, we would like to point out that the concept of combining skyrmions on various surfaces has been generalized in a number of publications. For instance, three skyrmion crystals can be entangled and stabilized using Monte Carlo simulations [140, 141]. However, because the topological charge for the magnetization in these objects is finite, they do not display the benefits of antiferromagnetic skyrmions.

### 3.2 Ferrimagnetic skyrmions

For a related entity, the ferrimagnetic skyrmion, signs of the favorable emerging electrodynamics of antiferromagnetic skyrmions have also been observed [142, 143]. Like the antiferromagnetic skyrmion, it consists of two linked subskyrmions with mutually reversible spins. Although the magnetic forces on the two sublattices are of different magnitudes, this results in an uncompensated magnetization, which made it possible to identify ferrimagnetic skyrmions in GdFeCo films using X-ray imaging [142]. There is a certain temperature where the skyrmion Hall effect does not exist when these particles are powered by spin currents [143]. Due to differing gyromagnetic ratios for the magnetic moments in the various sublattices at this temperature, the angular momentum is adjusted at this temperature even though the magnetization is not [143]. A reduced skyrmion Hall angle of $\theta$Sk = 20° has been seen at room temperature [142], but this full correction of the skyrmion Hall effect has yet to be observed experimentally. Furthermore, it has recently been experimentally discovered that domain walls driven by SOT may travel at speeds of up to 6 km/s in ferrimagnetic insulators close to the compensation temperature [144], suggesting that ferrimagnetic skyrmions may also be able to move at these speeds in the future.

Due to their indigent magnetization, ferrimagnetic skyrmions have the benefit of being easier to detect and interact with than antiferromagnetic ones. They promise similar benefits to antiferromagnetic skyrmions in terms of their emergent electrodynamics. However, a straight-line motion along a driving current is only anticipated to function at a specific temperature (angular momentum adjustment), which restricts the use of spintronics.

## 4. Applications of magnetic skyrmions

For information and communications technology, magnetic skyrmions may present a special opportunity to incorporate topology into room-temperature electronic systems. Although topological properties are present in many of the most fascinating recent developments in condensed matter physics, such as high-temperature interfacial superconductivity, the quantum Hall effect, or topological insulators, magnetic skyrmions take on a special significance because they are arguably the most promising candidates to be used in consumer-grade low-energy nanoscale spintronic devices in the medium term.

There are still many problems to be solved in the newly emerging field of study exploiting skyrmion topological features, as discussed in earlier parts. However, all of the fundamental operations of writing information (nucleation of individual skyrmions), processing information (displacement, creation/annihilation of skyrmions, excitation of skyrmion modes), and reading information (electrical detection of individual skyrmions) have already been individually demonstrated, though frequently only at low temperatures and for a skyrmion lattice rather than for a single skyrmion at room temperature.

The next major hurdle will be incorporating all three of these features into a single, tiny integrated device while operating at ambient temperature. Following, we briefly go through some of the more intriguing skyrmionic device concepts that have lately been put forth (for an extensive review of the application of skyrmions, we refer to W. Kang et al. [145].

### 4.1 Skyrmion race track memory

The first is referred to as the "Skyrmion racetrack memory," whose theory is quite similar to the one based on domain walls put forward by S. S. P. Parkin [5]. A series of individual skyrmions in a magnetic track can encode information by taking advantage of the solitonic nature of skyrmions [8, 146]. The great level of integration that skyrmions have over domain walls is one of their advantages [88]. By narrowing the track, the calculated diameter of a compact skyrmion can be reduced by several orders of magnitude, and as a result, in current-induced motion, the skyrmion accurately moves along the center of the track.

The distance between neighboring skyrmions in a track can be of the order of the skyrmion diameter, as simulations [88] have also demonstrated. As a result, one can anticipate a higher density with skyrmions than with domain walls in racetrack memory. Regarding energy consumption, despite the fact that the flexibility of the skyrmions' shape and/or trajectories should allow them to move with incredibly small depinning currents, as observed in skyrmion lattices in bulk materials [123], the ratio of velocity over current density is not expected to differ significantly between domain wall and skyrmions. Skyrmions have the additional benefit of being guided by the confinement of the track's edges, which makes it so that their motion by spin torques will be similar in both straight and curved sections of the track. The motion of domain walls, on the other hand, will be influenced by curved sections of the racetrack because torques in the wall behave differently in the inner and outer parts of the track. Finally, it is possible to count the number of skyrmions going over a track using conventional tunnel magneto-resistive devices, such as domain walls, or older methods that relied on unique transport signatures related to the topological character of skyrmions, such as the topological Hall effect. It is interesting how quickly a nanoscale voltage-gated skyrmion transistor can be created using this skyrmion racetrack concept. By including a gate in a specific section of the track, X. Zhang et al. [113] proposed this new function, which controls whether or not the skyrmion equivalent of a transistor's "on/off" switch passes. This new function locally modifies the magnetic properties of the magnetic medium, specifically the perpendicular anisotropy or the Dzyaloshinsky-Moriya interaction, by applying an electric field.

### 4.2 Skyrmionic logic devices

The possibility of a skyrmion functioning as a standalone "particle" has also led to the conceptualization of a number of spin logic devices based on skyrmions. The majority of them are based on results from micromagnetic simulations [147] performed in nanoscale wires of varying widths that demonstrate how a single skyrmion can be converted into a domain-wall pair and vice versa. By creating certain nanostructures, this conversion process theoretically enables basic logical operations such as the duplication or merger of skyrmions at will. Recently, X. Zhang et al. [148] developed skyrmion logic gates AND and OR based on these extra functionalities, realizing the first step toward a comprehensive logical architecture with the goal of surpassing the current spin logic devices, particularly in terms of their level of integration.

### 4.3 Skyrmion magnonic crystal

Skyrmions can be artificially arranged in a periodic pattern in a 1 dimensional or two-dimensional nanostructure by applying a local magnetic [147], electric [149, 150], or spin-polarized current [96] field, for example, locally applying a local electric [149, 151] or magnetic field, or injecting spin-polarized current [96]. The propagation of spin waves inside this innovative sort of "meta material" can then be tailored using such skyrmion lattices as a periodic modulation of the magnetization. Indeed, F. Ma et al. [151] recently demonstrated through numerical simulations that such skyrmion-based magnonic crystals have a significant advantage over more conventional ones (based on a periodic modulation of the magnetic properties induced typically by the lithography process) in that they can be dynamically reconfigured by simply adjusting the diameter of the skyrmions (by applying a magnetic field), altering the periodicity of the lattice, or even erasing it. Also take note of the fact that skyrmion crystals at the nanoscale scale are conceivable, although typical magnonic crystals made using current lithography techniques are not [96].

By utilizing this functionality, it should be possible to dynamically switch between the full rejection and full transmission of spin waves in a waveguide. The spin waves themselves may be in a topological phase while propagating in a two-dimensional atomic-size skyrmion lattice, which should enable the realization of the spin-wave counterpart of the anomalous quantum Hall effect for electrons [152].

### 4.4 Skyrmion-based radio frequency devices

The topological character of skyrmions may cause a disruptive step in nanoscale radio frequency devices, another class of component. The low-frequency breathing mode, for instance, is a dynamical mode of a single skyrmion in a dot that is representative of its topological nature [153]. It has been suggested [154] that if the skyrmion-containing dot is a component of a magneto-resistive device like a spin valve or a magnetic tunnel junction, the skyrmion breathing mode brought on by spin torques can be utilized to produce a radio frequency signal.

The fact that the resulting skyrmion-based spin torque oscillator is based on a localized soliton makes it less vulnerable to external perturbation and more likely to exhibit a coherent dynamic than, say, a spin torque oscillator based on a vortex. The concept of a skyrmion-based microwave detector, which relies on the resonant excitation of the breathing mode when the frequency of the external radio frequency signal equals the breathing mode's frequency (that can be significantly changed by the application of an external perpendicular field, for example), and the conversion of this resonant dynamics into a direct current mixed voltage, is another function numerically investigated by G. Finocchio et al. [155].

Finally, F. Garcia-Sanchez et al. [156] recently presented a novel sort of skyrmion-based spin torque oscillator that is based on the self-sustained gyration resulting from the competition between confinement from boundary edges and the spin forces owing to an inhomogeneous spin polarizer. There is no threshold current for the start of the skyrmion dynamics; hence, the corresponding gyro tropic frequency is about an order of magnitude lower than in typical vortex-based spin torque oscillators.

## 5. Conclusion

This review's summary has covered stability, emergent electrodynamics, associated alternative magnetic quasiparticles, and potential magnetic skyrmion-based applications. Fundamental excitations (both topologically trivial and non-trivial), excitations' variations, and extensions make up the diversity of particles. Skyrmions with an arbitrary helicity and antiskyrmions have been suggested as being the most technologically useful kinds of fundamental excitations. Topological excitations in a varied magnetic field and the fusion of numerous subparticles are two examples of variations. Antiferromagnetic and ferrimagnetic skyrmions have been thoroughly covered in this chapter. All of these things can be arranged in two dimensions (both periodically and ad hoc) and extended along a third dimension, which is what is meant by the term "extension." Bloch points, chiral bobbers, and other naturally three-dimensional objects have been the focus of our attention.

Antiferromagnetic skyrmions are frequently regarded as the best parts for spintronic applications out of the items that have been given. They can be driven by currents at extremely high speeds of up to several kilometers per second because of their corrected magnetic texture, and the absence of the Skyrmion Hall effect eliminates the issue of bit pinning at the margins. Additionally, due to their local adjustment of magnetization, stray fields are reduced, enabling the three-dimensional stacking of many racetracks in close proximity to one another.

Magnetic skyrmions may offer a unique chance for room-temperature electronic devices to incorporate topology in information and communications technologies. Magnetic skyrmions assume a special significance because they are arguably the most promising candidates to be used in consumer-grade low-energy nanoscale spintronic devices in the medium term. Topological properties are present in many of the most fascinating recent developments in condensed matter physics, such as high-temperature interfacial superconductivity, the quantum Hall effect, and topological insulators. Skyrmion racetrack memory, a skyrmion logical device, a skyrmion magnetic crystal, and skyrmion-based radio frequency devices are some of the numerous practical benefits based on magnetic skyrmions.

## Acknowledgements

We acknowledge Mrs. Bezawit Alemneh for typing the first draft of the chapter.

## Funding

There is no funding for this work.

## Conflict of interest

The author confirms that there is no conflict of interest.

## Data availability

All the data used to support the findings of this study are included in the chapter.

## Author details

Birhanu Abera Kolech
Department of Physics, Natural and Computational Science, Mekdela Amba University, Tulu Awuliya, Ethiopia

*Address all correspondence to: birhanuabera643@gmail.com

## References

[1] Moore GE. Cramming more components onto integrated circuits. IEEE Solid-state Circuits Society Newsletter. 1965;**38**(8):114

[2] Waldrop MM. The chips are down for Moore's law. Nature News. 2016; **530**(7589):144

[3] Parkin SS. International Business Machines Corp, assignee. Shiftable magnetic shift register and method of using the same. United States patent US 6,834,005. 2004

[4] Parkin SS, Hayashi M, Thomas L. Magnetic domain-wall racetrack memory. Science. 2008;**320**(5873): 190-194

[5] Parkin S, Yang SH. Memory on the racetrack. Nature Nanotechnology. 2015; **10**(3):195-198

[6] Bogdanov AN, Yablonskii DA. Thermodynamically stable "vortices" in magnetically ordered crystals. The mixed state of magnets. Zh. Eksp. Teor. Fiz. 1989;**95**(1):178

[7] Mühlbauer S, Binz B, Jonietz F, Pfleiderer C, Rosch A, Neubauer A, et al. Skyrmion lattice in a chiral magnet. Science. 2009;**323**(5916):915 919

[8] Sampaio J, Cros V, Rohart S, Thiaville A, Fert A. Nucleation, stability and current-induced motion of isolated magnetic skyrmions in nanostructures. Nature Nanotechnology. 2013;**8**(11): 839-844

[9] Fert A, Cros V, Sampaio J. Skyrmions on the track. Nature Nanotechnology. 2013;**8**(3):152-156

[10] Yu G, Upadhyaya P, Shao Q, Wu H, Yin G, Li X, et al. Room-temperature skyrmion shift device for memory application. Nano Letters. 2017;**17**(1): 261-268

[11] Hamamoto K, Ezawa M, Nagaosa N. Purely electrical detection of a skyrmion in constricted geometry. Applied Physics Letters. 2016;**108**(11): 112401

[12] Maccariello D, Legrand W, Reyren N, Garcia K, Bouzehouane K, Collin S, et al. Electrical detection of single magnetic skyrmions in metallic multilayers at room temperature. Nature Nanotechnology. 2018;**13**(3):233-237

[13] Sondhi SL, Karlhede A, Kivelson SA, Rezayi EH. Skyrmions and the crossover from the integer to fractional quantum Hall effect at small Zeeman energies. Physical Review B. 1993;**47**(24):16419

[14] Al Khawaja U, Stoof H. Skyrmions in a ferromagnetic Bose–Einstein condensate. Nature. Jun 2001;**411** (6840):918-920

[15] Skyrme TH. A non-linear field theory. Proceedings of the Royal Society of London. Series a. Mathematical and Physical Sciences. 1961;**260**(1300): 127-138

[16] Skyrme TH. A unified field theory of mesons and baryons. Nuclear Physics. 1962;**31**:556-569

[17] Finkelstein D, Rubinstein J. Connection between spin, statistics, and kinks. Journal of Mathematical Physics. 1968;**9**(11):1762-1779

[18] Adkins GS, Nappi CR, Witten E. Static properties of nucleons in the Skyrme model. Nuclear Physics B. 1983; **228**(3):552-566

[19] Vilenkin A, Shellard EP. Cosmic Strings and Other Topological Defects.

Cambridge: Cambridge University Press; 1994

[20] Fukuda JI, Žumer S. Quasi-two-dimensional Skyrmion lattices in a chiral nematic liquid crystal. Nature Communications. 2011;**2**(1):1-5

[21] Rózsa L, Palotás K, Deák A, Simon E, Yanes R, Udvardi L, et al. Formation and stability of metastable skyrmionic spin structures with various topologies in an ultrathin film. Physical Review B. 2017; **95**(9):094423

[22] Göbel B, Mertig I, Tretiakov OA. Beyond Skyrmions: Review and Perspectives of Alternative Magnetic Quasiparticles [Internet]. Vol. 895. Physics Reports. Elsevier BV; 2021. pp. 1-28. DOI: 10.1016/j.physrep.2020.10.001

[23] Yu XZ, Onose Y, Kanazawa N, Park JH, Han JH, Matsui Y, et al. Real-space observation of a two-dimensional skyrmion crystal. Nature. 2010; **465**(7300):901-904

[24] Bogdanov A, Hubert A. Thermodynamically stable magnetic vortex states in magnetic crystals. Journal of Magnetism and Magnetic Materials. 1994;**138**(3):255-269

[25] Nagaosa N, Tokura Y. Topological properties and dynamics of magnetic skyrmions. Nature Nanotechnology. 2013;**8**(12):899-911

[26] Zhou Y. Magnetic skyrmions: Intriguing physics and new spintronic device concepts. National Science Review. 2019;**6**(2):210-212

[27] Bobeck AH, Bonyhard PI, Geusic JE. Magnetic bubbles—An emerging new memory technology. Proceedings of the IEEE. 1975;**63**(8):1176-1195

[28] Davies J, Clover R, Lieberman B, Rose D. Reliability considerations in the design of one-megabit bubble memory chips. IEEE Transactions on Magnetics. 1980;**16**(5):1106-1110

[29] Grünberg P, Barnas J, Saurenbach F, Fuβ JA, Wolf A, Vohl M. Layered magnetic structures: Antiferromagnetic type interlayer coupling and magnetoresistance due to antiparallel alignment. Journal of magnetism and magnetic materials. 1991 Feb 1;**93**:58-66

[30] Binash G, Gruuberg P, Saurenbach F. Enhanced magnetoresistance in layered magnetic interlayers exchange giant magnetoresistance. Physical Review B. 1989;**39**:4828-4834

[31] Masuoka F, Momodomi M, Iwata Y, Shirota R. New ultra high density EPROM and flash EEPROM with NAND structure cell. In: 1987 International Electron Devices Meeting. Kawasaki: VLSI research center, IEEE; 1987. pp. 552-555

[32] Malinowski G, Boulle O, Kläui M. Current-induced domain wall motion in nanoscale ferromagnetic elements. Journal of Physics D: Applied Physics. 2011;**44**(38):384005

[33] Tretiakov OA, Tchernyshyov O. Vortices in thin ferromagnetic films and the skyrmion number. Physical Review B. 2007;**75**(1):012408

[34] Kooy C. Experimental and theoretical study of the domain configuration in thin layers of BaFe_ < 12> O_< 19. Philosophical Research Reports. 1960;**15**(7):1180-1186

[35] Thiele AA. Steady-state motion of magnetic domains. Physical Review Letters. 1973;**30**(6):230

[36] Malozemoff AP, Slonczewski JC. Magnetic Domain Walls in Bubble Materials. New York: Academic Press; 1979

[37] Heinze S, Von Bergmann K, Menzel M, Brede J, Kubetzka A, Wiesendanger R, et al. Spontaneous atomic-scale magnetic skyrmion lattice in two dimensions. Nature Physics. 2011; 7(9):713-718

[38] Garlow JA, Pollard SD, Beleggia M, Dutta T, Yang H, Zhu Y. Quantification of mixed Bloch-Néel topological spin textures stabilized by the Dzyaloshinskii-Moriya interaction in Co/ Pd multilayers. Physical Review Letters. 2019;**122**(23):237201

[39] Nayak AK, Kumar V, Ma T, Werner P, Pippel E, Sahoo R, et al. Magnetic antiskyrmions above room temperature in tetragonal Heusler materials. Nature. 2017;**548**(7669):561-566

[40] Leonov AO, Mostovoy M. Multiply periodic states and isolated skyrmions in an anisotropic frustrated magnet. Nature Communications. 2015;**6**(1):1-8

[41] Neubauer A, Pfleiderer C, Binz B, Rosch A, Ritz R, Niklowitz PG, et al. Topological Hall effect a phase of MnSi. Physical Review Letters. 2009;**102**(18): 186602

[42] Adams T, Mühlbauer S, Pfleiderer C, Jonietz F, Bauer A, Neubauer A, et al. Long-range crystalline nature of the skyrmion lattice in MnSi. Physical Review Letters. 2011;**107**(21):217206

[43] Mühlbauer S, Kindervater J, Adams T, Bauer A, Keiderling U, Pfleiderer C. Kinetic small angle neutron scattering of the skyrmion lattice in MnSi. New Journal of Physics. 2016; **18**(7):075017

[44] Lee M, Kang W, Onose Y, Tokura Y, Ong NP. Unusual Hall effect anomaly in MnSi under pressure. Physical Review Letters. 2009;**102**(18):186601

[45] Ritz R, Halder M, Franz C, Bauer A, Wagner M, Bamler R, et al. Giant generic topological Hall resistivity of MnSi under pressure. Physical Review B. 2013; **87**(13):134424

[46] Chacon A, Bauer A, Adams T, Rucker F, Brandl G, Georgii R, et al. Uniaxial pressure dependence of magnetic order in MnSi. Physical Review Letters. 2015;**115**(26):267202

[47] Nii Y, Nakajima T, Kikkawa A, Yamasaki Y, Ohishi K, Suzuki J, et al. Uniaxial stress control of skyrmion phase. Nature Communications. 2015; **6**(1):1-7

[48] Tonomura A, Yu X, Yanagisawa K, Matsuda T, Onose Y, Kanazawa N, et al. Real-space observation of skyrmion lattice in helimagnet MnSi thin samples. Nano Letters. 2012;**12**(3):1673-1677

[49] Yu X, Kikkawa A, Morikawa D, Shibata K, Tokunaga Y, Taguchi Y, et al. Variation of skyrmion forms and their stability in MnSi thin plates. Physical Review B. 2015;**91**(5):054411

[50] Adams T, Mühlbauer S, Neubauer A, Münzer W, Jonietz F, Georgii R, et al. Skyrmion lattice domains in Fe1− xCoxSi. Journal of Physics: Conference Series. 2010;**200**(3):032001

[51] Münzer W, Neubauer A, Adams T, Mühlbauer S, Franz C, Jonietz F, et al. Skyrmion lattice in the doped semiconductor Fe 1− x Co x Si. Physical Review B. 2010;**81**(4):041203

[52] Huang SX, Chien CL. Extended skyrmion phase in epitaxial FeGe (111) thin films. Physical Review Letters. 2012; **108**(26):267201

[53] McGrouther D, Lamb RJ, Krajnak M, McFadzean S, McVitie S, Stamps RL, et al. Internal structure of hexagonal

skyrmion lattices in cubic helimagnets. New Journal of Physics. 2016;**18**(9): 095004

[54] Yu XZ, Kanazawa N, Onose Y, Kimoto K, Zhang WZ, Ishiwata S, et al. Near room-temperature formation of a skyrmion crystal in thin-films of the helimagnet FeGe. Nature Materials. 2011;**10**(2):106-109

[55] Adams T, Chacon A, Wagner M, Bauer A, Brandl G, Pedersen B, et al. Long-wavelength helimagnetic order and skyrmion lattice phase in Cu 2 OSeO 3. Physical Review Letters. 2012;**108**(23): 237204

[56] Seki S, Yu XZ, Ishiwata S, Tokura Y. Observation of skyrmions in a multiferroic material. Science. 2012; **336**(6078):198-201

[57] Okamura Y, Kagawa F, Seki S, Tokura Y. Transition to and from the skyrmion lattice phase by electric fields in a magnetoelectric compound. Nature Communications. 2016;**7**(1):1-6

[58] Zhang SL, Bauer A, Berger H, Pfleiderer C, Van Der Laan G, Hesjedal T. Resonant elastic x-ray scattering from the skyrmion lattice in Cu 2 OSeO 3. Physical Review B. 2016; **93**(21):214420

[59] Zhang SL, Bauer A, Berger H, Pfleiderer C, Van Der Laan G, Hesjedal T. Imaging and manipulation of skyrmion lattice domains in Cu2OSeO3. Applied Physics Letters. 2016;**109**(19): 192406

[60] Zhang SL, Bauer A, Burn DM, Milde P, Neuber E, Eng LM, et al. Multidomain skyrmion lattice state in Cu2OSeO3. Nano Letters. 2016;**16**(5): 3285-3291

[61] Rajeswari J, Huang P, Mancini GF, Murooka Y, Latychevskaia T, McGrouther D, et al. Filming the formation and fluctuation of skyrmion domains by cryo-Lorentz transmission electron microscopy. Proceedings of the National Academy of Sciences. 2015; **112**(46):14212-14217

[62] Karube K, White JS, Reynolds N, Gavilano JL, Oike H, Kikkawa A, et al. Robust metastable skyrmions and their triangular–square lattice structural transition in a high-temperature chiral magnet. Nature Materials. 2016;**15**(12): 1237-1242

[63] Tokunaga Y, Yu XZ, White JS, Rønnow HM, Morikawa D, Taguchi Y, et al. A new class of chiral materials hosting magnetic skyrmions beyond room temperature. Nature Communications. 2015;**6**(1):1-7

[64] Kézsmárki I, Bordács S, Milde P, Neuber E, Eng LM, White JS, et al. Néel-type skyrmion lattice with confined orientation in the polar magnetic semiconductor GaV4S8. Nature Materials. 2015;**14**(11):1116-1122

[65] von Bergmann K, Kubetzka A, Pietzsch O, Wiesendanger R. Interface-induced chiral domain walls, spin spirals and skyrmions revealed by spin-polarized scanning tunneling microscopy. Journal of Physics. Condensed Matter. 2014;**26**(39):394002

[66] Wiesendanger R. Nanoscale magnetic skyrmions in metallic films and multilayers: A new twist for spintronics. Nature Reviews Materials. 2016;**1**(7):1-1

[67] Romming N, Hanneken C, Menzel M, Bickel JE, Wolter B, von Bergmann K, et al. Writing and deleting single magnetic skyrmions. Science. 2013;**341**(6146):636-639

[68] Romming N, Kubetzka A, Hanneken C, von Bergmann K,

Wiesendanger R. Field-dependent size and shape of single magnetic skyrmions. Physical Review Letters. 2015;**114**(17): 177203

[69] Schmidt L, Hagemeister J, Hsu PJ, Kubetzka A, Von Bergmann K, Wiesendanger R. Symmetry breaking in spin spirals and skyrmions by in-plane and canted magnetic fields. New Journal of Physics. 2016;**18**(7):075007

[70] Hanneken C, Kubetzka A, Von Bergmann K, Wiesendanger R. Pinning and movement of individual nanoscale magnetic skyrmions via defects. New Journal of Physics. 2016;**18**(5): 055009

[71] Moreau-Luchaire C, Moutafis C, Reyren N, Sampaio J, Vaz CA, Van Horne N, et al. Additive interfacial chiral interaction in multilayers for stabilization of small individual skyrmions at room temperature. Nature Nanotechnology. 2016;**11**(5):444-448

[72] Seki S, Ishiwata S, Tokura Y. Magnetoelectric nature of skyrmions in a chiral magnetic insulator Cu 2 OSeO 3. Physical Review B. 2012;**86**(6):060403

[73] White JS, Levatić I, Omrani AA, Egetenmeyer N, Prša K, Živković I, et al. Electric field control of the skyrmion lattice in Cu2OSeO3. Journal of Physics. Condensed Matter. 2012;**24**(43):432201

[74] Milde P, Neuber E, Bauer A, Pfleiderer C, Berger H, Eng LM. Heuristic description of magnetoelectricity of Cu2OSeO3. Nano Letters. 2016;**16**(9):5612-5618

[75] Onose Y, Okamura Y, Seki S, Ishiwata S, Tokura Y. Observation of magnetic excitations of skyrmion crystal in a helimagnetic insulator Cu 2 OSeO 3. Physical Review Letters. 2012;**109**(3): 037603

[76] Mochizuki M. Spin-wave modes and their intense excitation effects in skyrmion crystals. Physical Review Letters. 2012;**108**(1):017601

[77] Schwarze T, Waizner J, Garst M, Bauer A, Stasinopoulos I, Berger H, et al. Universal helimagnon and skyrmion excitations in metallic, semiconducting and insulating chiral magnets. Nature Materials. 2015 May;**14**(5):478-483

[78] Kugler M, Brandl G, Waizner J, Janoschek M, Georgii R, Bauer A, et al. Band structure of helimagnons in MnSi resolved by inelastic neutron scattering. Physical Review Letters. 2015;**115**(9): 097203

[79] Nii Y, Kikkawa A, Taguchi Y, Tokura Y, Iwasa Y. Elastic stiffness of a skyrmion crystal. Physical Review Letters. 2014;**113**(26):267203

[80] Kruchkov AJ, Rønnow HM. Skyrmion lattices in electric fields. arXiv preprint arXiv:1702.08863. 2017

[81] Buhrandt S, Fritz L. Skyrmion lattice phase in three-dimensional chiral magnets from Monte Carlo simulations. Physical Review B. 2013; **88**(19):195137

[82] Nagaosa N, Yu XZ, Tokura Y. Gauge fields in real and momentum spaces in magnets: Monopoles and skyrmions. Philosophical Transactions of the Royal Society A - Mathematical Physical and Engineering Sciences. 2012;**370**(1981): 5806-5819

[83] Schütte C, Rosch A. Dynamics and energetics of emergent magnetic monopoles in chiral magnets. Physical Review B. 2014;**90**(17):174432

[84] Schütte C. Skyrmions and monopoles in chiral magnets &

correlated heterostructures (Doctoral dissertation, Universität zu Köln).

[85] Milde P, Köhler D, Seidel J, Eng LM, Bauer A, Chacon A, et al. Unwinding of a skyrmion lattice by magnetic monopoles. Science. 2013; **340**(6136):1076-1080

[86] Bogdanov A. New localized solutions of the nonlinear field equations. Jetp Letters. 1995;**62**(3):247-251

[87] Lin SZ, Reichhardt C, Batista CD, Saxena A. Particle model for skyrmions in metallic chiral magnets: Dynamics, pinning, and creep. Physical Review B. 2013;**87**(21):214419

[88] Zhang X, Zhao GP, Fangohr H, Liu JP, Xia WX, Xia J, et al. Skyrmion-skyrmion and skyrmion-edge repulsions in skyrmion-based racetrack memory. Scientific Reports. 2015;**5**(1):1-6

[89] Rózsa L, Deák A, Simon E, Yanes R, Udvardi L, Szunyogh L, et al. Skyrmions with attractive interactions in an ultrathin magnetic film. Physical Review Letters. 2016;**117**(15):157205

[90] Leonov AO, Kézsmárki I. Asymmetric isolated skyrmions in polar magnets with easy-plane anisotropy. Physical Review B. 2017;**96**(1):014423

[91] Leonov AO, Kézsmárki I. Skyrmion robustness in noncentrosymmetric magnets with axial symmetry: The role of anisotropy and tilted magnetic fields. Physical Review B. 2017;**96**(21):214413

[92] Leonov AO, Monchesky TL, Loudon JC, Bogdanov AN. Three-dimensional chiral skyrmions with attractive interparticle interactions. Journal of Physics: Condensed Matter. 2016;**28**(35):35LT01

[93] Leonov AO, Loudon JC, Bogdanov AN. Spintronics via non-axisymmetric chiral skyrmions. Applied Physics Letters. 2016;**109**(17):172404

[94] Purnama I, Gan WL, Wong DW, Lew WS. Guided current-induced skyrmion motion in 1D potential well. Scientific Reports. 2015;**5**(1):1-9

[95] Iwasaki J, Mochizuki M, Nagaosa N. Universal current-velocity relation of skyrmion motion in chiral magnets. Nature Communications. 2013;**4**(1):1-8

[96] Iwasaki J, Mochizuki M, Nagaosa N. Current-induced skyrmion dynamics in constricted geometries. Nature Nanotechnology. 2013;**8**(10): 742-747

[97] Schütte C, Iwasaki J, Rosch A, Nagaosa N. Inertia, diffusion, and dynamics of a driven skyrmion. Physical Review B. 2014;**90**(17):174434

[98] Everschor-Sitte K, Sitte M. Real-space Berry phases: Skyrmion soccer. Journal of Applied Physics. 2014;**115**(17): 172602

[99] Makhfudz I, Krüger B, Tchernyshyov O. Inertia and chiral edge modes of a skyrmion magnetic bubble. Physical Review Letters. 2012;**109**(21): 217201

[100] Büttner F, Moutafis C, Schneider M, Krüger B, Günther CM, Geilhufe J, et al. Dynamics and inertia of skyrmionic spin structures. Nature Physics. 2015;**11**(3):225-228

[101] Kong L, Zang J. Dynamics of an insulating skyrmion under a temperature gradient. Physical Review Letters. 2013; **111**(6):067203

[102] Mochizuki M, Yu XZ, Seki S, Kanazawa N, Koshibae W, Zang J, et al. Thermally driven ratchet motion of a

skyrmion microcrystal and topological magnon Hall effect. Nature Materials. 2014;**13**(3):241-246

[103] Iwasaki J, Beekman AJ, Nagaosa N. Theory of magnon-skyrmion scattering in chiral magnets. Physical Review B. 2014;**89**(6):064412

[104] Lin SZ, Batista CD, Saxena A. Internal modes of a skyrmion in the ferromagnetic state of chiral magnets. Physical Review B. 2014;**89**(2):024415

[105] Schütte C, Garst M. Magnon-skyrmion scattering in chiral magnets. Physical Review B. 2014;**90**(9): 094423

[106] Schroeter S, Garst M. Scattering of high-energy magnons off a magnetic skyrmion. Low Temperature Physics. 2015;**41**(10):817-825

[107] Zhang X, Ezawa M, Xiao D, Zhao GP, Liu Y, Zhou Y. All-magnetic control of skyrmions in nanowires by a spin wave. Nanotechnology. 2015; **26**(22):225701

[108] Garcia-Sanchez F, Borys P, Vansteenkiste A, Kim JV, Stamps RL. Nonreciprocal spin-wave channeling along textures driven by the Dzyaloshinskii-Moriya interaction. Physical Review B. 2014;**89**(22): 224408

[109] Seki S, Okamura Y, Kondou K, Shibata K, Kubota M, Takagi R, et al. Magnetochiral nonreciprocity of volume spin wave propagation in chiral-lattice ferromagnets. Physical Review B. 2016; **93**(23):235131

[110] Wang W, Beg M, Zhang B, Kuch W, Fangohr H. Driving magnetic skyrmions with microwave fields. Physical Review B. 2015;**92**(2): 020403

[111] Kim JV, Yoo MW. Current-driven skyrmion dynamics in disordered films. Applied Physics Letters. 2017;**110**(13): 132404

[112] Jiang W, Zhang X, Yu G, Zhang W, Wang X, Benjamin Jungfleisch M, et al. Direct observation of the skyrmion Hall effect. Nature Physics. 2017;**13**(2): 162-169

[113] Zhang X, Zhou Y, Ezawa M, Zhao GP, Zhao W. Magnetic skyrmion transistor: Skyrmion motion in a voltage-gated nanotrack. Scientific Reports. 2015;**5**(1):1-8

[114] Zhang X, Zhou Y, Ezawa M. Magnetic bilayer-skyrmions without skyrmion Hall effect. Nature Communications. 2016;**7**(1):1-7

[115] Rodrigues DR, Everschor-Sitte K, Tretiakov OA, Sinova J, Abanov A. Spin texture motion in antiferromagnetic and ferromagnetic nanowires. Physical Review B. 2017;**95**(17):174408

[116] Hall EH. On a new action of the magnet on electric currents. American Journal of Science. 1880;**3**(111):200-205

[117] Smit J, Volger J. Spontaneous Hall effect in ferromagnetics. Physical Review. 1953;**92**(6):1576

[118] Smit J. The spontaneous Hall effect in ferromagnetics II. Physica. 1958;**24**(1–5):39-51

[119] Berry MV. Quantal phase factors accompanying adiabatic changes. Proceedings of the Royal Society of London. A. Mathematical and Physical Sciences. 1984;**392**(1802):45-57

[120] Taguchi Y, Oohara Y, Yoshizawa H, Nagaosa N, Tokura Y. Spin chirality, Berry phase, and anomalous Hall effect

in a frustrated ferromagnet. Science. 2001;**291**(5513):2573-2576

[121] Binz B, Vishwanath A, Aji V. Theory of the helical spin crystal: A candidate for the partially ordered state of MnSi. Physical Review Letters. 2006; **96**(20):207202

[122] Zang J, Mostovoy M, Han JH, Nagaosa N. Dynamics of skyrmion crystals in metallic thin films. Physical Review Letters. 2011;**107**(13):136804

[123] Yu XZ, Kanazawa N, Zhang WZ, Nagai T, Hara T, Kimoto K, et al. Skyrmion flow near room temperature in an ultralow current density. Nature Communications. 2012;**3**(1):1-6

[124] Lakshmanan M. The fascinating world of the Landau–Lifshitz–Gilbert equation: An overview. Philosophical Transactions of the Royal Society A - Mathematical Physical and Engineering Sciences. 2011;**369**(1939):1280-1300

[125] Barker J, Tretiakov OA. Static and dynamical properties of antiferromagnetic skyrmions in the presence of applied current and temperature. Physical Review Letters. 2016;**116**(14):147203

[126] Zhang X, Zhou Y, Ezawa M. Antiferromagnetic skyrmion: Stability, creation and manipulation. Scientific Reports. 2016;**6**(1):1-8

[127] Göbel B, Mook A, Henk J, Mertig I. Antiferromagnetic skyrmion crystals: Generation, topological Hall, and topological spin Hall effect. Physical Review B. 2017;**96**(6):060406

[128] Tveten EG, Qaiumzadeh A, Tretiakov OA, Brataas A. Staggered dynamics in antiferromagnets by collective coordinates. Physical Review Letters. 2013;**110**(12):127208

[129] Jin C, Song C, Wang J, Liu Q. Dynamics of antiferromagnetic skyrmion driven by the spin Hall effect. Applied Physics Letters. 2016;**109**(18): 182404

[130] Bessarab PF, Yudin D, Gulevich DR, Wadley P, Titov M, Tretiakov OA. Stability and lifetime of antiferromagnetic skyrmions. Physical Review B. 2019;**99**(14):140411

[131] Legrand W, Maccariello D, Ajejas F, Collin S, Vecchiola A, Bouzehouane K, et al. Room-temperature stabilization of antiferromagnetic skyrmions in synthetic antiferromagnets. Nature Materials. 2020;**19**(1):34-42

[132] Dohi T, DuttaGupta S, Fukami S, Ohno H. Formation and current-induced motion of synthetic antiferromagnetic skyrmion bubbles. Nature Communications. 2019;**10**(1):1-6

[133] Jiang W, Upadhyaya P, Zhang W, Yu G, Jungfleisch MB, Fradin FY, et al. Blowing magnetic skyrmion bubbles. Science. 2015;**349**(6245):283-286

[134] Liang X, Xia J, Zhang X, Ezawa M, Tretiakov OA, Liu X, et al. Antiferromagnetic skyrmion-based logic gates controlled by electric currents and fields. Applied Physics Letters. 2021; **119**(6):062403

[135] Everschor-Sitte K, Masell J, Reeve RM, Kläui M. Perspective: Magnetic skyrmions—Overview of recent progress in an active research field. Journal of Applied Physics. 2018; **124**(24):240901

[136] Buhl PM, Freimuth F, Blügel S, Mokrousov Y. Topological spin Hall effect in antiferromagnetic skyrmions. (RRL)–Rapid Research Letters. 2017; **11**(4):1700007

[137] Akosa CA, Tretiakov OA, Tatara G, Manchon A. Theory of the topological spin Hall effect in antiferromagnetic skyrmions: Impact on current-induced motion. Physical Review Letters. 2018; **121**(9):097204

[138] Cheng R, Niu Q. Electron dynamics in slowly varying antiferromagnetic texture. Physical Review B. 2012;**86**(24): 245118

[139] Gomonay O. Berry-phase effects and electronic dynamics in a noncollinear antiferromagnetic texture. Physical Review B. 2015;**91**(14):144421

[140] Rosales HD, Cabra DC, Pujol P. Three-sublattice skyrmion crystal in the antiferromagnetic triangular lattice. Physical Review B. 2015;**92**(21):214439

[141] Díaz SA, Klinovaja J, Loss D. Topological magnons and edge states in antiferromagnetic skyrmion crystals. Physical Review Letters. 2019;**122**(18): 187203

[142] Woo S, Song KM, Zhang X, Zhou Y, Ezawa M, Liu X, et al. Current-driven dynamics and inhibition of the skyrmion Hall effect of ferrimagnetic skyrmions in GdFeCo films. Nature Communications. 2018;**9**(1):1-8

[143] Kim SK, Lee KJ, Tserkovnyak Y. Self-focusing skyrmion racetracks in ferrimagnets. Physical Review B. 2017; **95**(14):140404

[144] Zhou HA, Dong Y, Xu T, Xu K, Sánchez-Tejerina L, Zhao L, Ba Y, Gargiani P, Valvidares M, Zhao Y, Carpentieri M. Compensated magnetic insulators for extremely fast spin-orbitronics. arXiv preprint arXiv: 1912.01775. 2019

[145] Kang W, Huang Y, Zhang X, Zhou Y, Zhao W. Skyrmion-electronics: An overview and outlook. Proceedings of the IEEE. 2016;**104**(10):2040-2061

[146] Kang W, Huang Y, Zheng C, Lv W, Lei N, Zhang Y, et al. Voltage controlled magnetic skyrmion motion for racetrack memory. Scientific Reports. 2016;**6**(1): 1-1

[147] Zhou Y, Ezawa M. A reversible conversion between a skyrmion and a domain-wall pair in a junction geometry. Nature Communications. 2014;**5**(1):1-8

[148] Zhang X, Ezawa M, Zhou Y. Magnetic skyrmion logic gates: Conversion, duplication and merging of skyrmions. Scientific Reports. 2015;**5**(1): 1-8

[149] Hsu PJ, Kubetzka A, Finco A, Romming N, Von Bergmann K, Wiesendanger R. Electric-field-driven switching of individual magnetic skyrmions. Nature Nanotechnology. 2017;**12**(2):123-126

[150] Schott M, Bernand-Mantel A, Ranno L, Pizzini S, Vogel J, Béa H, et al. The skyrmion switch: Turning magnetic skyrmion bubbles on and off with an electric field. Nano Letters. 2017;**17**(5): 3006-3012

[151] Ma F, Zhou Y, Braun HB, Lew WS. Skyrmion based dynamic magnonic crystal. Nano Letters. 2015;**15**(6): 4029-4036

[152] Roldán-Molina A, Nunez AS, Fernández-Rossier J. Topological spin waves in the atomic-scale magnetic skyrmion crystal. New Journal of Physics 2016;**18**(4):045015

[153] Kim JV, Garcia-Sanchez F, Sampaio J, Moreau-Luchaire C, Cros V, Fert A. Breathing modes of confined skyrmions in ultrathin magnetic dots. Physical Review B. 2014;**90**(6):064410

[154] Carpentieri M, Tomasello R, Zivieri R, Finocchio G. Topological, non-topological and instanton droplets driven by spin-transfer torque in materials with perpendicular magnetic anisotropy and Dzyaloshinskii–Moriya interaction. Scientific Reports. 2015;**5**(1):1-8

[155] Finocchio G, Ricci M, Tomasello R, Giordano A, Lanuzza M, Puliafito V, et al. Skyrmion based microwave detectors and harvesting. Applied Physics Letters. 2015;**107**(26):262401

[156] Garcia-Sanchez F, Sampaio J, Reyren N, Cros V, Kim JV. A skyrmion-based spin-torque nano-oscillator. New Journal of Physics. 2016;**18**(7):075011

Chapter 4

# Identification of Trailing Vortex Dynamic States

*Anthony P. Brown*

## Abstract

Flight research of the characteristics of trailing vortices, generated by heavy jet transport aircraft in cruising flight, was conducted. Trailing vortex velocities were derived by vectorial differencing of aircraft inertial velocity and true airspeed vectors, and then transforming to the vortex mean axis. Lateral distances between port and starboard vortices were 60–70% of generator wingspan. Vortex core radii were derived. Core pressure states were expanded or diffused. Core diffusion was associated with axial segmentation. Core pressure expansions included magnitudes greater than Euler equilibrium values, with velocity profiles displaying peaked maxima. Associated with these characteristics was vortex core radial instability. Subsequent radial expansion and contraction resulted in a large range of $r_C$ values. Vorticity confined to an annular state and discretized into circular arrays of N-point vortices of small $r_C$ was prevalent. Radial profiles of vortex velocity were identified and included Rankine (peaked) profiles, Lamb-Oseen, and Burnham-Hallock rounded profiles. Twenty-five percent of identified profiles were rounded. The majority of profiles were peaked, with maxima greater than, or equal to, Rankine values. Temperature gradients inside and outside of core edges were identified: outside, heating occurred, inside, cooling. Outer heating occurred with upstream axial flow. Inner cooling occurred with downstream axial flow.

**Keywords:** wake turbulence, wake vortex, vortex profile identification, vortex necklaces, circular N-point vortices, core diffusion, vortex tube heating, axial flow

## 1. Introduction

The Flight Research Laboratory of the NRC Canada has undertaken a large number of projects concerning enroute jet transport aircraft wake turbulence flight research. Critically, enroute jet transport aircraft wake turbulence is invariably characterized by a pair of trailing, contra-rotating line vortices. Following formation by the shedding and rolling-up of lift-inducing wing-bound vorticity, at a spacing governed by the lateral distribution of wing lift, the line vortices descend, in a momentum balance with wing lift, and transport themselves laterally, in coupled behavior to background atmospheric shear and stratification. Each vortex induces flow velocities on its counterpart. For example, the tangential velocity will



IntechOpen

be accelerated on the vortex pair inner sides and decelerated on the outer sides. This so-called mutual shortwave or elliptical instability could destabilize the vortex state of the pair, moreso if coupled to the negative axial pressure gradient, extending from the baseflow region behind the jet aircraft generating the trailing vortices. In addition, mutual attraction between vortices could excite a longwave or Crow instability, when coupling to the turbulence structure of the background atmosphere results in amplification of the instability mode. That amplification could develop, to the point of linking of the port and starboard vortices, followed by self-propulsion of the resultant vortex rings, or, conversely, segmentation, fragmentation and demise.

This summary of trailing vortex characteristics was formed during, and subsequent to, the initial research flights of the late 1960s, early 1970s [1]. The flight research of that period included that of the NRC, using a CT-133 military jet trainer, to measure the trailing vortex characteristics from DC-8 and CV-880 wake generators [2]. This flight research included the discovery of smaller, more intense vortex elements, within the basic trailing vortex pair flow-field. Their existence was not necessarily commensurate with the simple models of shortwave and longwave instability, described above.

In 2004/5, the NRC re-instrumented the CT-133 with advanced NRC inertial, airdata and data acquisition systems, capable of measuring unsteady winds at rates of 600 samples per second. Cruising flight wake turbulence flight research was recommenced in 2004 [3], initially using the NRC Falcon, and continued, using the NRC CT-133 over a fifteen year period, from 2006 [4]. The research flights measured trailing vortex state, lateral separation, lateral and vertical locations. It emphasized vortex state, discovered the spectra of core radii, $r_C$, states in spatiotemporal segments of the trailing vortices, without any significant change to mean $r_C$ values, over the ages of the organized trailing vortices [5]. The flight measurements enabled the identification of vortex core radial instability, vortex velocity, pressure, temperature, vorticity and circulation profiles, together with axial flow characteristics. Spatiotemporal venting and diffusion of vortex core pressure was prevalent, resulting in vorticity purging to unstable annular distributions, resulting in discretization to rings of N-point vortices and subsequent reformation [6]. Additionally, longwave instability was identified as invariably multi-wavelength.

## 2. Experimental details

### 2.1 Measurement aircraft and wake generators

The NRC CT-133 research aircraft (**Figure 1**) undertook flight measurements. Fitted with specialized NRC inertial and air data instrumentation systems (**Figure 1** close-up), the aircraft acquired inertial and air data at 600 Hertz (Hz), thereby measuring the instantaneous inertial and true air speed (TAS) vectors. Wake generator aircraft types were A310, 333, 343, 346, 359, 388 and B744, 752, 763, 772/3, 788/9, where 'A' designates Airbus, 'B' designates Boeing, the first two numerals designate type and the third designates variant, e.g. A359 is an A350-900).

The TAS vector (i.e. relative to the aircraft) was then transformed through the aircraft Euler angles of pitch, roll and yaw, to earth axes. Vectorial differencing thence derived the instantaneous wind vector. Subtracting mean winds left the vortex-induced air flow velocity vector, whence in the vicinity of wake vortices.

**Figure 1.**
*NRC. CT-133 research aircraft; close-up view of air data nose-boom, including pitot-static tube, and cylinders in cross-flow for flow angle measurement.*

## 2.2 Flight profiles and wake vortex measurement

Wake vortex research flights were conducted from Ottawa Airport. Heavy Category jet transports flying in the vicinity were intercepted, at normal regulatory separation, with the assistance of air traffic control [3]. When 9–10 km behind, the CT-133 was flown downwards or upwards between the pair of trailing vortices, then around each individually. In this manner the net circulation of the pair was derived by open path integrals referenced to the plane orthogonal to the vortex axis (i.e. from infinity to the joining line between vortices, subtended angle π/2 to each vortex), $\int_O^P \tilde{V}.ds = \int_O^P \frac{\Gamma}{2\pi(r_m/Cos\theta)}.Cos\theta\frac{r_m d\theta}{Cos^2\theta} = \int_{-\pi/2}^{0}\frac{\Gamma}{2\pi}d\theta = \Gamma/4$, for each vortex of the pair, (**Figure 2**), so that for the pair of vortices of opposite sign,, $\int_0^Q V\cdot ds = (|\Gamma_L|-|\Gamma_R|)/2$ yields the mean modulus of circulation; the gross circulation of each vortex was given by closed path integral, $\Gamma = \oint_C V \cdot ds = \iint_S n(\nabla \mathbf{x} V).dS = \iint_S \omega\cdot dS$ (also **Figure 2**) [7].

Examples of the derivation is shown in **Figure 3**, for the cases of enroute encounters with B744 and A388 wake vortex flow-fields. In the B744 case, the encounter geometry and wake vortex state were unknown. Nevertheless the open path integral defined the temporality of the encounter, whilst the unsteady wind vector reversals, defined the vortex core edge- encounters. The A388 vortex locations were accurately measured.

Following flight between the vortex pair and flight around each vortex, individual vortex cores were traversed, entering the cores from below. Vortex core rotation generally loaded and rolled the CT-133 inwards, towards the centreplane of the pair. **Figure 4** depicts the CT-133 video imagery of the vortex core traverse of an A388 vortex.

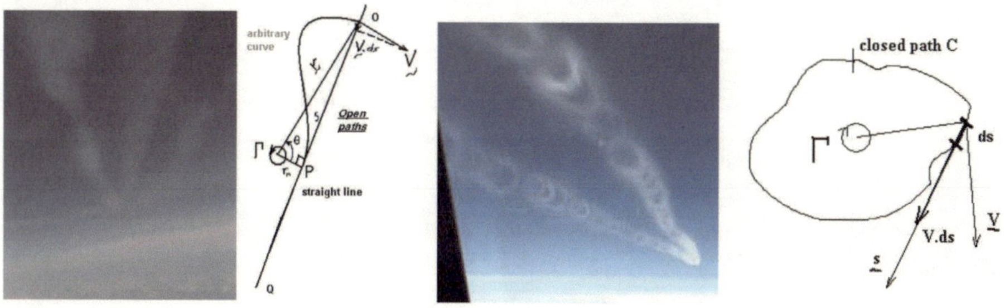


**Figure 2.**
*Flight path wake vortex views: (left), A343 vortices at 0.6 min age during climb (and upper jet wake condensate between), (left centre), open path integral for a curved approach flight path from above, (centre right) B763 vortex pair at 1 min age, and closed path integral for individual vortex circulation [7].*

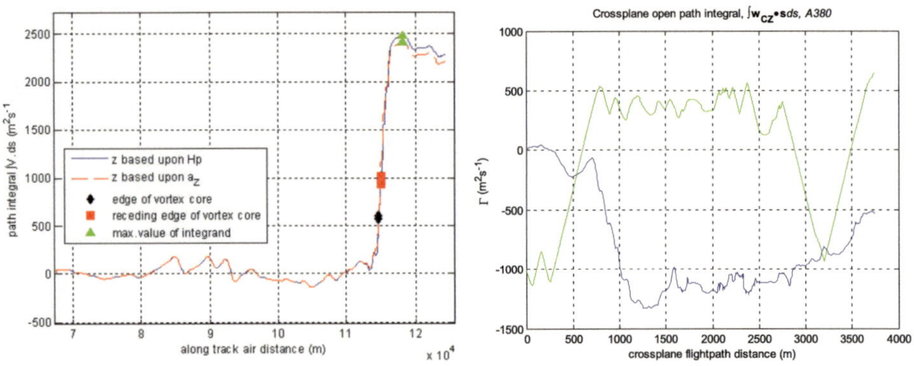


**Figure 3.**
*Open path integrals, referenced to the vortex crossplane, of the winds experienced by WVE aircraft: An Airbus A319 encounter with a Boeing 747-400 (B744) wake (left) [8]; NRC CT-133 climb (altitude in green) and encounter (open path circulation in blue) with an Airbus A380-800 (A388) trailing vortex pair (right).*

## 3. Results and discussion

### 3.1 Vortex pair tilting, interaction with atmospheric shear

The tilting of the vortex plane, that containing the port and starboard vortices (parallel to the wing plane at vortex formation behind the wing, hence horizontal for a jet transport aircraft in cruise) is observed in **Figure 4**, at an angle of $\approx 30°$. Possibly the principal cause of vortex plane tilting was highlighted by Mokry [10] to be interaction with background atmospheric vertical shear. Tilting is accompanied by upper vortex strengthening [10], lower vortex weakening, accompanied by diminishment of condensate density, due to a reduction in the core pressure expansion, associated with the weakening. Hence the net circulation of the vortex pair will no longer be zero. The phenomenon was highlighted by initial Falcon measurements [3, 11]. The plot of **Figure 5** shows quantified values of net circulation plotted against background

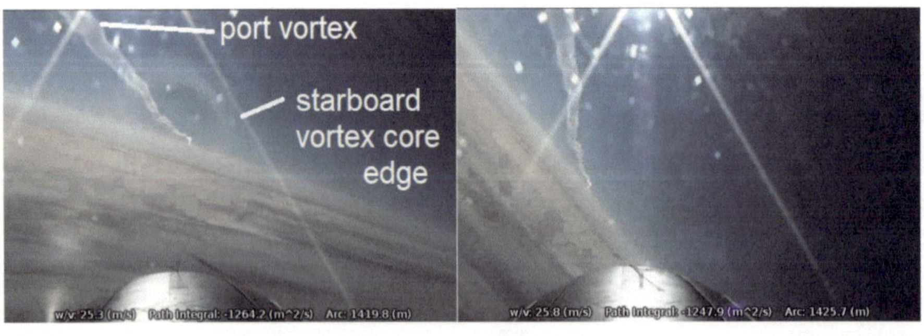


**Figure 4.**
*vortex core entry (left) and exit (right) of the curved-inwards, near-vertical traverse through the starboard vortex core of an A388 aircraft, of duration 0.7 sec. Overlaid data indicated core edge vortex velocity was 26 m/s, circulation as per* ***Figure 3****, crossplane flight path distance was 6 m; circular reconstruction of the core identified a core radius of 3.6 m [9].*

atmospheric shear. For the Boeing 767–300 (B763) case, the net circulation in moderate shear was 38 ± 13% that of initially generated circulation, implying strengthening and weakening of generated vortices, 19 ± 7% each.

### 3.2 Vortex core circulation distribution

For the estimation of circulation from vortex core traverses, CT-133 flight-path assumptions have been made; in particular, that velocity spatial gradients in the vortex axial (forward flight) direction are at least an order of magnitude less than those in the crossplane (i.e. orthogonal to the vortex axis), a reasonable assumption for line vortices, such as aircraft trailing vortices. Hence, the essentially longitudinal flight-path can be concatenated in the crossplane, in which 'quasi-vorticity' between sequential [y z] points, having vortex velocity components of $[w_y\ w_z]$ has been formulated [12] as

$$o_{yz} = \partial w_y/\partial z - \partial w_z/\partial y \approx (w_{yi+1} - w_{yi})/(z_{i+1} - z_i) - (w_{zi+1} - w_{zi})/(y_{i+1} - y_i) \quad (1)$$

For the derivation of circulation from $o_{yz}$ by area integration (**Figure 6**), a further assumption was required, namely one-dimensionality (radial) of $o_{yz}$ distributions. Consider this assumption: discussed earlier, mutual induction of each vortex on the other will invariably induce elliptical (shortwave) instability. The magnitude of such instability varies inversely to the ratio of lateral distance between vortex centres, divided by core radius. Given that the flight data yielded an $r_C$ independence of wingspan, then circularity and one-dimensionality were reasonable approximations for large (Heavy Category) jet transports.

Thence, circulation was derived by Green's Theorem, as $\Gamma = \iint \omega_{yz}.dydz$. The distribution of vortex velocity, quasi-vorticity and circular integration thereof are shown in **Figure 6** for a number of vortex core vertical traverses, all from the same wake [12]. Each traverse was undertaken in an elapsed time of 0.7 ± 0.3 s, in which duration, the CT-133 flew 170 ± 40 m in the vortex axial direction. Therefore, when approaching and receding flight-path segments were concatenated into radial plots from derived vortex centres, the overlays of **Figure 6** revealed any changes of vortex core state in the axial direction.

As implied by the samples in **Figure 6**, vortex profiles were spatiotemporally-variant. The variations between rounded BH profiles, peaked Rankine profiles and

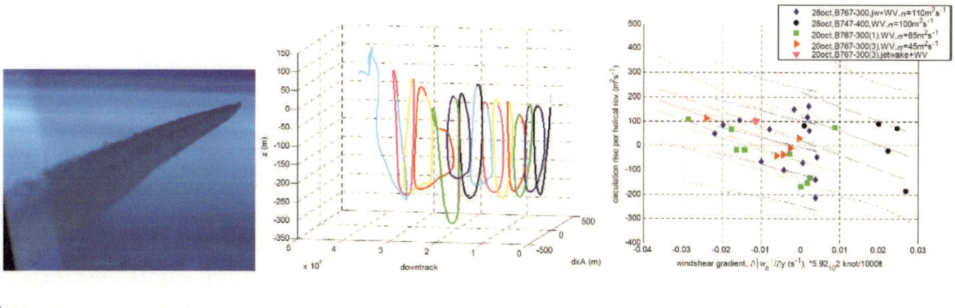


**Figure 5.**
*Analysis of net circulation of vortex pairs in background crossplane/vertical windshear: (left) B763 in unusually dense condensate (upper jet wake condensate observed; vortices are buried within the lower region of the condensate): (centre) Falcon flight-path around the overall condensate field; (right), plot of net circulation against vertical shear.*

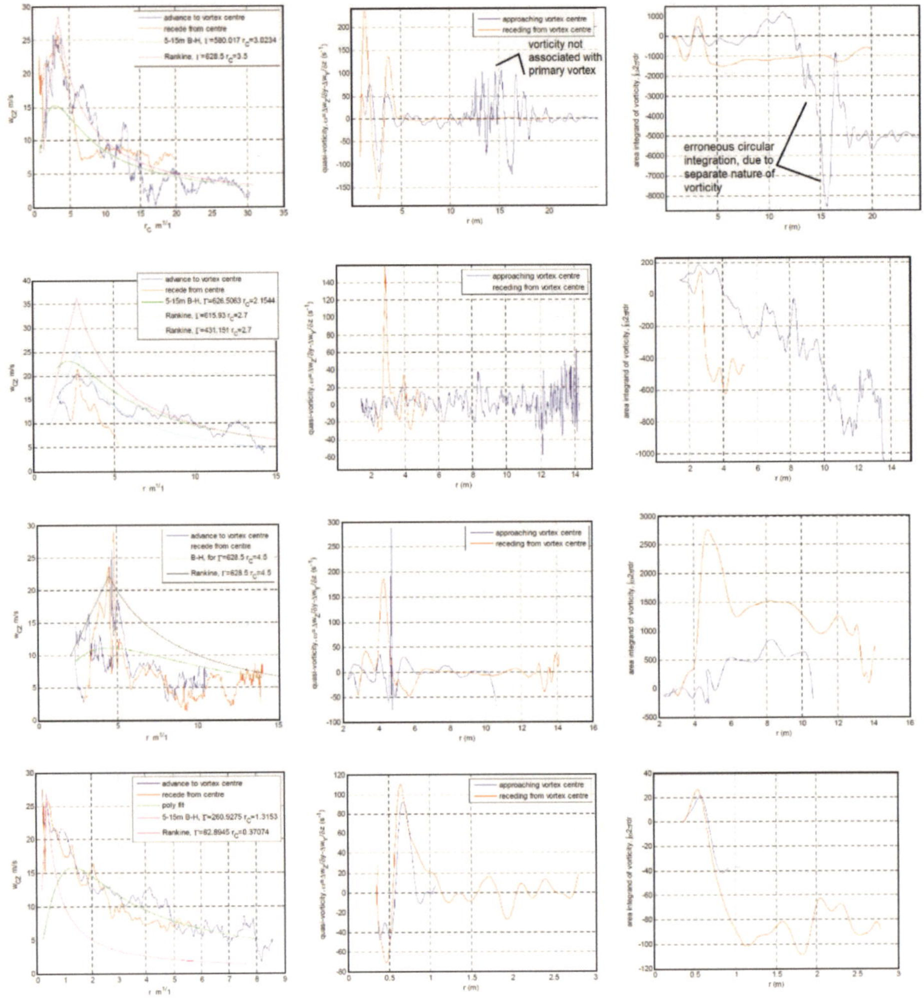


**Figure 6.**
*Radial distributions of (left) vortex velocities, (centre) $o_{yz}$, (right) $\omega 2\pi r o_{yz} dr$, four vortex core vertical traverses from the aircraft: wake age 50–80 sec., generated $\Gamma$ = 735 $m^2/s$: (top) traverse of* ***Figure 4****, fair symmetry of approach/recessive peaked profiles, approximated by the Rankine profile; (middle), differing approach and recessive profiles, similar velocity maxima, 20 m/s, better approximated by the Burnham Hallock (BH) profile; (lower middle) highly peaked profile, due to vorticity concentration at core edge; (bottom), small radius vortex core state, very symmetrical between approach and recession, much smaller circulation.*

core-edge peaked profiles were generally present, regardless of the type of wake generating aircraft. Also, **Figure 6** intimated the presence of opposite sign vorticity inside or outside of core edges, from which area-integration would result in non-monotonic rises of circulation, with increasing radius. Saffmann [13] considered it likely that trailing vortex structures would include the presence of Taylor instabilities, over-circulation and relaxation in the outer vortex radius. With regard to the evidence of axial direction spatiotemporal variations in vortex profile state, circular integration of $o_{yz}$ would yield significant error in apparent circulation—indeed the case in **Figure 6**, although, qualitatively, 'over-circulation' and radial relaxation therefrom, is observed in cases.

### 3.3 Vortex lateral separation

The measured values of lateral separation between port and starboard vortex centres $b_V$, are shown in **Figure 7**, as functions of wake generator wingspan. For any particular wingspan value, variations in lateral separation between vortices was indicative of the longwave instability magnitude (linking of vortices would occur near zero lateral separation). Generally, the spacing is 0.6–0.7 b; 60 m wingspan data is skewed by long-wave excitation prevalence for that particular wake state. 0.6–0.7 b is a lower spacing that the $\pi/4$ = 0.79 b optimum spacing for an elliptical lift distribution for minimum induced drag. The lower spacing might have a been a swept wing effect at the high cruise Mach (generally 0.8–0.84) of the wake generators, at which outboard loading is reduced, due to swept attachment line boundary layer migration and shock-boundary layer interactions, resulting in greater inboard loading and lower $b_{V_o}$ at generation.

### 3.4 Vortex core radius

With such axial direction spatiotemporal variations in vortex profiles, variations in $r_C$ could also be expected, in any particular wake survey (**Figure 8** shows such variations). Examination of the abscissa dimensions in **Figure 8** supports this hypothesis. Derived $r_C$ values from several hundred core traverses at wake ages of 50–200 s, are presented in **Figure 8**, in dimensional and non-dimensional forms, normalized by wingspan of the wake generator, and by wake age, respectively [5, 14].

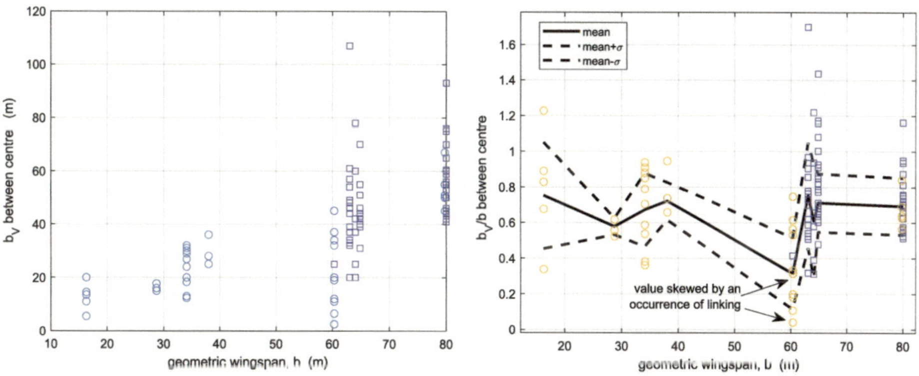


**Figure 7.**
*Flight data of measured values of lateral separation between port and starboard vortex centres $b_V$ (left), and normalized by wake generator geometric wingspan, $b_V/b$.*

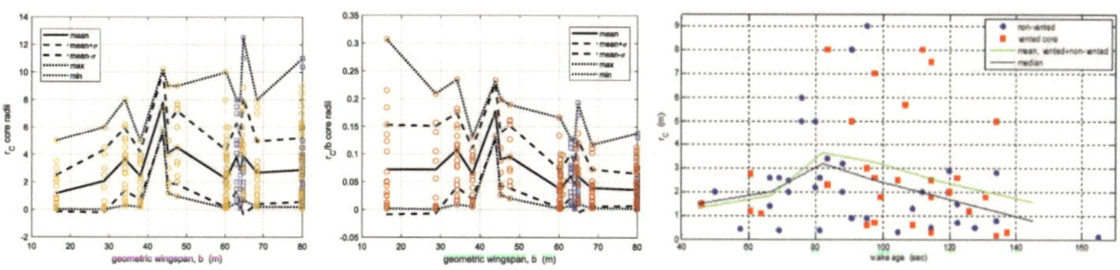

**Figure 8.**
*Trailing vortex core radius, variation of $r_C$ with wake generator wingspan, b (left), variation of $r_C/b$ with b (middle), and variation of $r_C$ with wake vortex age (right).*

Between the limits of the wingspan domain (16–80 m), there was a non-monotonic doubling of mean $r_C$ values, from 1.2 to 2.8 m, so that mean $r_C/b$ reduced from 0.07 to 0.04. However, for wingspans between 30 and 50 m, mean $r_C$ values were in the range of 4–8 m (10–17% b). Minima $r_C$ values were $\approx 0.06 \pm 0.04$ m, and did not vary with b. Plotted against wake age, mean and median $r_C$ showed an increase from 1.5 m at 48 s to 3 m at 80 s, then reduced to 1 m at 150 s age.

### 3.5 Vented vortex cores

**Figure 8** displays vortex core traverses as non-vented or vented. A vented core is one, in which the core pressure, $\Delta P_S$, is diffused between the core edge and centreline, raising $\Delta P_S$ partially or fully, to the background atmospheric pressure. Examples of vented and non-vented core data are shown in **Figure 9**.

When not vented, solution to the Euler equation for a vortex core [16], will result in a pressure distribution of further expansion inside the core edge. For both Rankine and BH profiles, the solution is centreline expansion, $\Delta P_{CL}$, being twice that of core edges, $\Delta P_{rc}$, (**Figure 10**). For a symmetric annular distribution of vorticity, area-integrating to the same $\Gamma$ (400 $m^2/s$ in this example), the machine-solved analytic

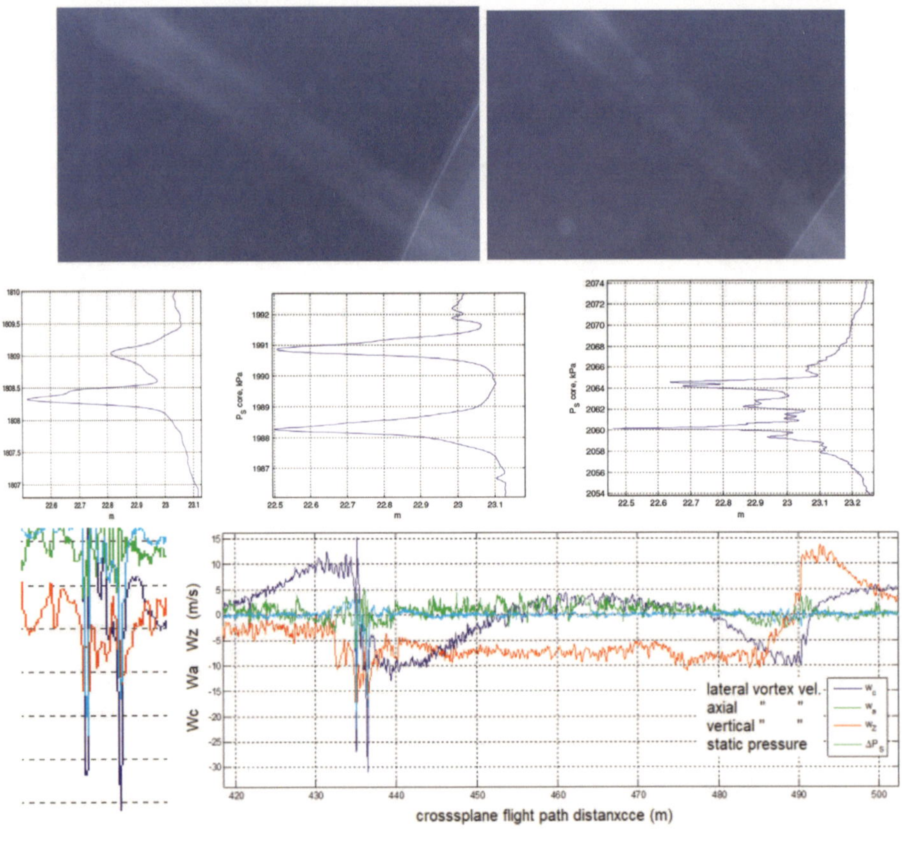


**Figure 9.**
*Examples of non-vented and vented trailing vortex core traverses: (upper) photos of DC-8 vortices at 18 km length, left, and 9 km, right; (mid) DC-8 vortex core diffusion, with progressively increasing wake length to 28 km [15]; (bottom right), B744 trailing pair traverse, port vortex vented (inset view shown bottom left), with downstream (−) axial flow, $w_A$, starboard vortex, unvented.*

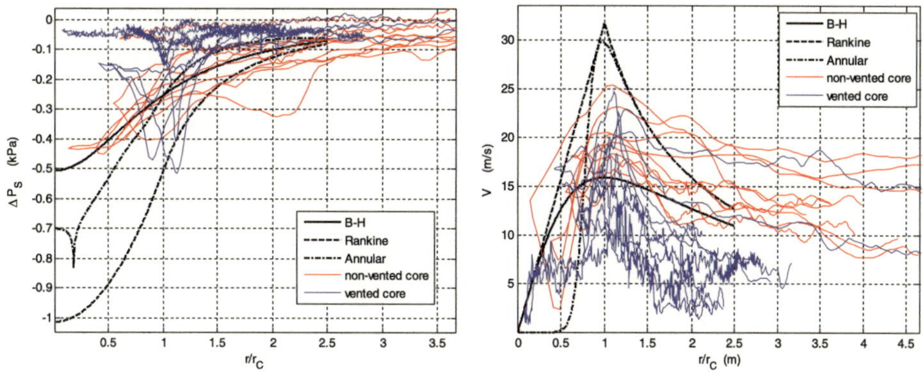


**Figure 10.**
*Comparison of analytical solutions to flight data examples (all from a single A359 wake vortex survey) of: (left), vortex core pressure distributions for Rankine, BH, and annular vorticity vortex core, & (right), vortex velocity distributions; against r/r$_C$.*

ratio of $\Delta P_{CL}/\Delta P_{rc}$, was 2.8, also shown in **Figure 10**. However, if the trailing vortices were to become axially segmented for example (i.e. 'open-ended') hydrostatically-induced pressure relaxation would occur spatiotemporally along the axis. Also shown in **Figure 10**, are the respective vortex velocity distributions. The analytical distributions of pressure and velocity are compared in the figure, for five non-vented core traverses and five vented core traverses.

Concerning $\Delta P_S$, it is seen that the non-vented cores (in red) have lower magnitudes of core edge expansion—one traverse follows the BH $\Delta P_S$ distribution very closely, including an inside-core inflection approaching $\Delta P_{CL}$. $\Delta P_{rc}$ for vented cores was similar or greater than the solution for Rankine profiles, but had high inside gradients of diffusion approaching ambient pressure by 0.7–0.8 $r_C$. Flight data core velocity distributions are seen to have had vortex velocity maxima, $V_{MAX}$, between BH and Rankine values. Non-vented velocity profiles and magnitudes were evocative of Lam-Oseen profiles, whereas vented profiles relaxed with large spatial gradients, to be ≈25% core edge maxima by ≈0.6$r_C$. The analytic solution for annular vorticity had a similar relaxation gradient.

By reference to **Figure 8**, it would appear that vented cores occurred throughout wake surveys, with greater frequency for ages greater than peak $r_C$ (≈80 s age, or 1.6 non-dimensional units of wake age). $\Delta P_{CL}/\Delta P_{rc}$ for a wide range and number of core traverses has been plotted against wake age in **Figure 11**. For non-vented core state, $\Delta P_{CL}/\Delta P_{rc}$ clustered around a value of ≈2;. Vented, the diffusion on/near the vortex centreline increased with age (diffusion magnitude enveloped by the brown line).

### 3.6 Statistical review of velocity profile shape and maxima

It has been noted above, that both peaked and rounded vortex profiles have appeared in trailing vortex traverse flight data. The profile measurements have been reviewed, statistically. Firstly, vortex velocity maxima have been analyzed. Considered against b (**Figure 12**), no particular relationship was apparent; against $r_C$, an asymptotic relationship was evident. Next, $V_{MAX}$ and profile shape have been considered together, in the assemblage of core traverses, wherefore radial distance has been normalized by $r_C$ for each traverse, **Figure 13a** [17]. Reference Rankine, BH [18] and FJ [19] profiles are overlaid on the flight measurements. Overall mean, mean ±σ, and

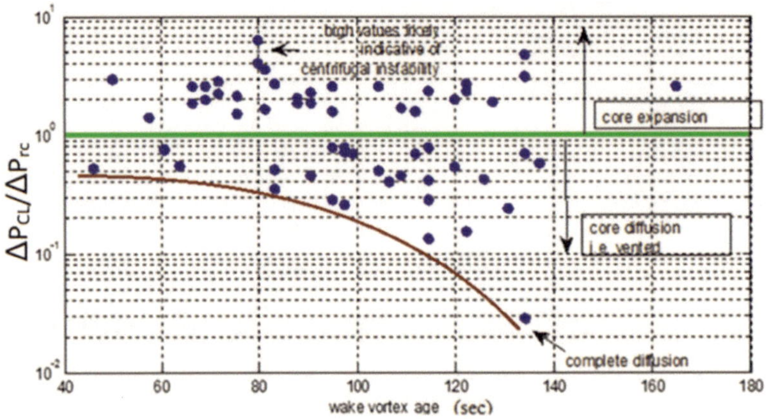


**Figure 11.**
*Vortex core traverse $\Delta P_{CL}/\Delta P_{rc}$ as a function of wake age.*

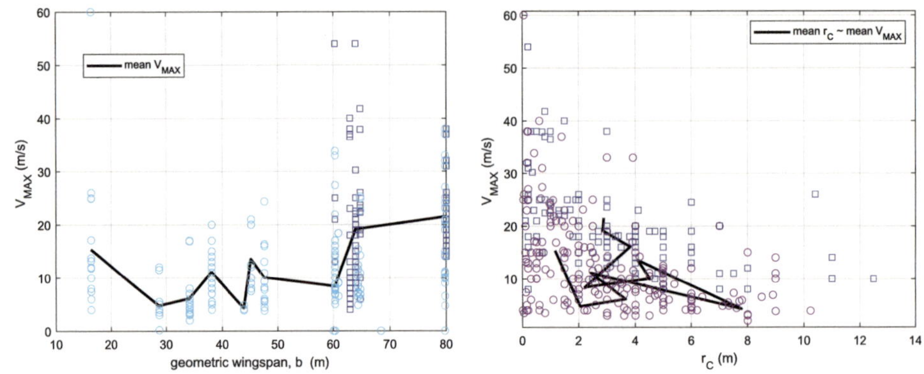


**Figure 12.**
*Vortex velocity maxima: left, plotted against b, with the mean $V_{MAX}$ values for the data clusters at each wingspan value; right, plotted against $r_C$—the black line is the locus plot of mean $V_{MAX}$ and mean $r_C$ values for each wingspan data cluster.*

mean $\pm 2\sigma$ of flight data profiles have been included in **Figure 13a**, all of which are peaked, rather than rounded. 25% of flight profile $V_{MAX}$ was enveloped by the BH peak magnitude, 50% lay between BH and Rankine ($V_{MAX}/V_{MAX\text{-}BH} = 2$) and the remaining 25% lay between Rankine and FJ (the latter, for $V_{MAX}/V_{MAX\text{-}BH} = 2.5$). The overall mean $V_{MAX}$ value was $1.5V_{MAX\text{-}BH}$.

### 3.7 Radial instability

Also shown in **Figure 13b** is the vortex radial stability parameter, for flight data statistical profiles and for the Rankine, BH and FJ profiles. Rankine [17] conducted vortex stability analysis, with the conclusion that, if $\Omega(r) = V(r)/r$, then the vortex will be radially stable if $d^2(r^2\Omega)^2/dr^2 > 0$, and conversely unstable if $<0$. Denoting the parameter as *Ra*, it was estimated for individual flight data profiles, by double numerical differencing, and of the statistical ensemble profiles of *mean, mean + σ, mean + 2σ*, whereas it can be derived analytically for the Rankine, BH and FJ profiles. Inside the core edge, for some individual core profiles $Ra < 0$, indicating instability.

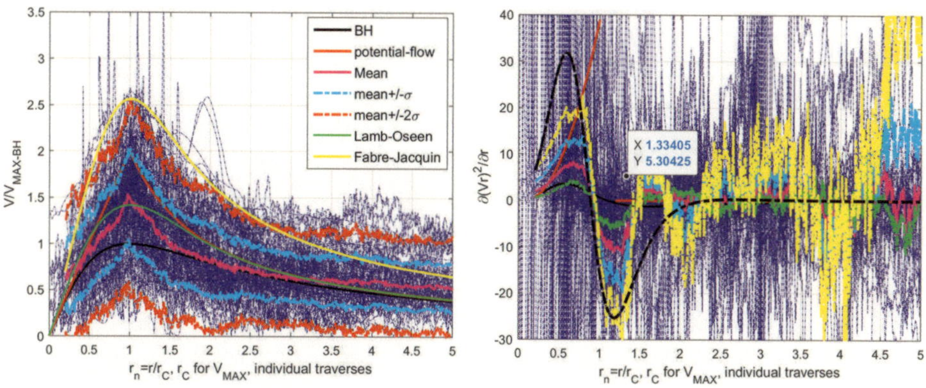


**Figure 13.**
*(a),* left, *statistical analysis of flight data vortex profiles & comparison with various vortex profile models, and (b),* right, *radial stability parameter analysis [17].*

Double numerical differencing of the statistical ensemble profiles and vortex models indicated *Ra* > 0 in the cores. Approaching core edges, for the statistical ensembles, stability reversal occurred, and *Ra* < 0 for all outside the core edge. The highest $V_{MAX}$ profile (mean + 2σ) had the greatest magnitude of radial instability, between $1 < r/r_C < 1.5$ and, further, $3.5 < r/r_C < 4.2$, beyond which a high level of radial stability was exhibited.

To investigate the temporality of such unstable *Ra* < 0 values for the *mean + 2σ* flight data profile, with maximum tangential velocity $V_\theta = 1.58(\Gamma/2\pi r_C)$, of **Figure 13**, numerical studies were conducted [20]. For this, the 3D (radial, axial directions and time, i.e. axisymmetric) set of Euler equations in cylindrical coordinates, pertinent to line vortex flow from [13],

$$\frac{\partial V_r}{\partial t} + V_r\frac{\partial V_r}{\partial r} - \frac{V_\theta^2}{r} + V_z\frac{\partial V_r}{\partial z} = -\frac{1}{\rho}\frac{\partial P}{\partial r}$$

$$\frac{\partial V_\theta}{\partial t} + V_r\frac{\partial V_\theta}{\partial r} + \frac{V_r V_\theta}{r} + V_z\frac{\partial V_\theta}{\partial z} = 0 \quad \text{with the continuity equation,}$$

$$\frac{\partial V_z}{\partial t} + V_r\frac{\partial V_z}{\partial r} + V_z\frac{\partial V_z}{\partial z} = -\frac{1}{\rho}\frac{\partial P}{\partial z}$$

$$\frac{1}{r}\frac{\partial (rV_r)}{\partial r} + \frac{\partial V_z}{\partial z} = 0, \tag{2}$$

and initial conditions over the [r z] domain, $V_z = V_r = 0$, $V_\theta = 1.58\Gamma/2\pi r^{5/4}$, and p(r) given by $p - p_\infty = \rho o(V_\theta^2 r)dry$. A numerical time-marching solution to this system of equations for $\Gamma = 200\ m^2/s$, $r_C = 1$ m and $V_\theta = 1.58\Gamma/2\pi r^{5/4}$ at high altitude, with $\rho = 0.3119\ kg/m^3$ yielded the solution as shown in **Figure 14**. [20], from the initial conditions of zero radial flow and tangential profile as prescribed.

In the first time-step, a strong radial outflow had onset at and beyond the vortex core edge. The radial outflow then propagated outwards (with $V_\theta$ subsiding for $r > r_C$, and by t = 0.14472 s, peak $V_\theta$ had established at $0.2r_C$), with an inwards flow established inside the core. The inwards flow propagated outwards. A reversal in $V_\theta$ direction occurred inside the core, between the new $V_\theta$ peak at $0.2r_C$ and $r_C$. $V_r$ and $V_\theta$ perturbations subsided with increasing radial distance, both being about 50% undisturbed $V_\theta$ magnitude at $4r_C$. Inboard, peak $V_r$ values were ± 40% of undisturbed

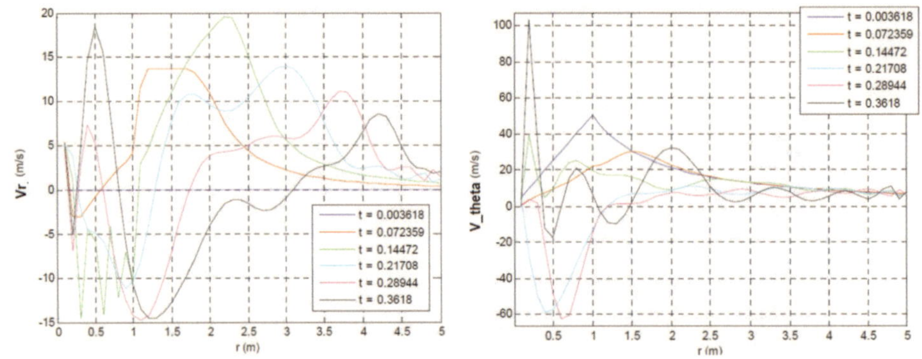


**Figure 14.**
*Time-step marching numerical solution, for $r_C = 1$ m, to the Euler equation set for line vortex flow: radial distributions of: radial* (left) *and tangential velocities* (right).

peak $V_\theta$, whilst transient $V_\theta$ values inside the original core were $\pm$ 100% of peak undisturbed $V_\theta$.

With such high values of transient $V_\theta$ in particular, and associated high shear in the $\theta$ direction, it would be expected to see smaller vortices, and associated vorticity and velocity variations, occur in the $\theta$ direction. Such variations of course would not be modeled with a set of axisymmetric equations. Nevertheless the example has been qualitatively instructive in the extent of core flow instability and $r_C$ variations that could be possible with high peaked vortex profiles, such as those frequently measured in aircraft trailing vortex traverses.

## 3.8 Comparison of vortex profiles to the Rankine profile

Another approach to the statistical analysis of vortex profiles derived from flight traverses is to compare peak $V_V$ values, $V_{MAX}$, to that which would occur in a Rankine profile for the same circulation and core radius, i.e. the parameter $V_{cf} = V_{MAX}/(\Gamma/2\pi r_C)$, wherefore $\Gamma \pm 10\%$ was derived from velocity profile fitting, or from generated $\Gamma$ minus an assumed loss proportional to age, where profile fitting was unsuccessful. $V_{cf}$ was plotted for the set of core traverses, against b and $r_C$, in **Figure 15**. *Mean* $V_{cf}$ and *mean* + $\sigma$ $V_{cf}$ against b were included, whilst *mean*_$V_{cf} \sim$ *mean*_$r_C$ for each wingspan, was also plotted. Plotted against b, the mean $V_{cf}$ values are seen to be approximately 1 for b < ≈50 m, and 0.5–0.8, for b > ≈50 m. When plotted against *mean* $r_C$ for each wingspan, *mean* $V_{cf}$ is seen to have been ≈1 for $1 < r_C < 8$ m, but to have had a bifurcating branch for $r_C < 4$ m, to a value of ≈0.25 at $r_C \approx 2$ m.

The bifurcated branch in particular, was further examined by plotting $1/V_{cf}$ against $r_C$ in **Figure 16**. The linear plot highlighted the bifurcation in $1/V_{cf}$, emanating from *mean* $V_{cf} = 1$ at *mean* $r_C < 4$ m, to a value of $1/V_{cf} = 4$ at $r_C = 2$ m, and an additional bifurcation, from the same origin, to $1/V_{cf} = 2$ at $r_C = 3$ m. A value of $1/V_{cf} = 2$ is indicative of a BH profile. However $1/V_{cf} = 4$ implied a different vortex state: two possible states could be, *either* a transient, 'deflated' vortex velocity, such as that of **Figure 14**, in an unstable radial outflow state, *or* a vortex element, which has a much lower circulation that the trailing vortex system. The latter vortex elemental state implied the existence of a state consisting of a number of vortices constituting the overall trailing vortex circulation.

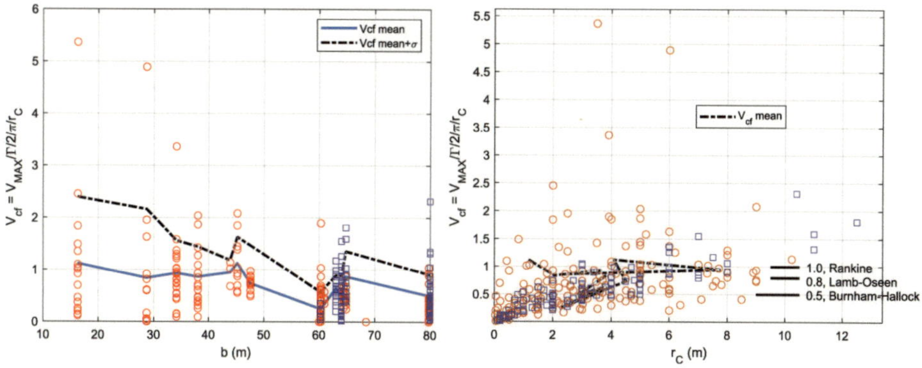


**Figure 15.**
*$V_{cf} = V_{MAX}/(\Gamma/2\pi r_C)$, plotted (left) against b, including mean and mean + σ $V_{cf}$; plotted (right) against $r_C$ for each core traverse and mean_$V_{cf}$ for each wingspan.*

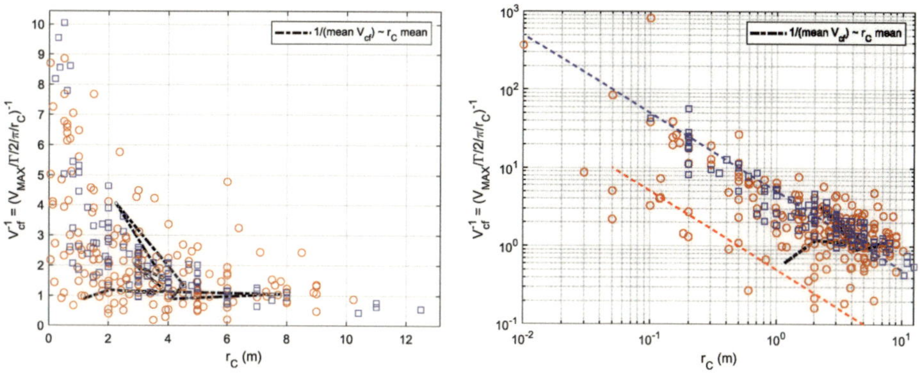


**Figure 16.**
*$1/V_{cf} \sim r_C$, (left) linear and (right) logarithmic plots.*

The logarithmic plot highlighted the inverse relationship of the $1/V_{cf}$ scaling with reducing $r_C$. The inverse scaling was bistatic, at $1/V_{cf} = 5/r_C$ and $1/V_{cf} = 0.5/r_C$.

## 3.9 Vented vortex core annularity

Vented vortex cores displayed partial or full diffusion of pressure on the vortex centreline (**Figures 9** and **10**), implying negligible vorticity near the centreline and migration/concentration of vorticity near the core edges; hence, an annular vorticity state. The annular state of trailing vortices shall be considered by reference to a specific example. **Figure 17** presents radial distributions of vortex velocity and pressure from a vortex core traverse [21]. The closest point of approach (CPA) to the reconstructed vortex centre was approximately $0.6r_C$. Direct pressure measurements indicated that the core was fully diffused by the CPA radial distance.

Concerning velocity profiles, the outer profile was close to that of a BH model. However, peak velocity magnitudes were much greater: at core-exit, $V_V$ was very close to that of the Rankine profile; at core-entry, it was 1.5 times that of the Rankine profile, similar to that of the stability study, **Figure 14**, [20], and well approximated by a profile of $V_V = \Gamma/2\pi r^3$. Possible annular (hollow) vortex models that could be

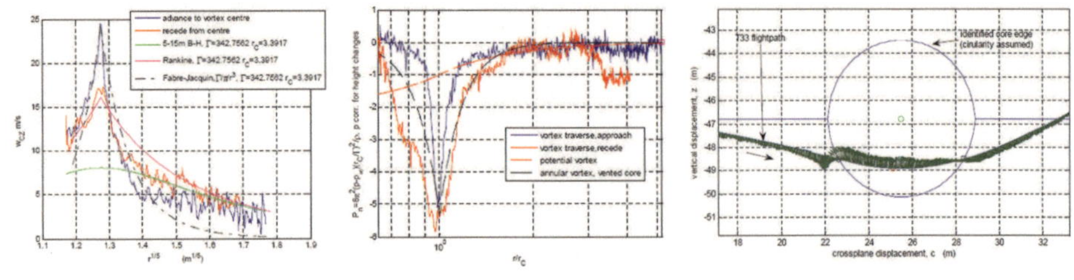

**Figure 17.**
*A388 vortex core traverse, wake age approximately 70 s (non-dimensional age 1.5, radial distributions: (left) vortex velocity, (centre) pressure, (right) placement of the flightpath across the vortex core.*

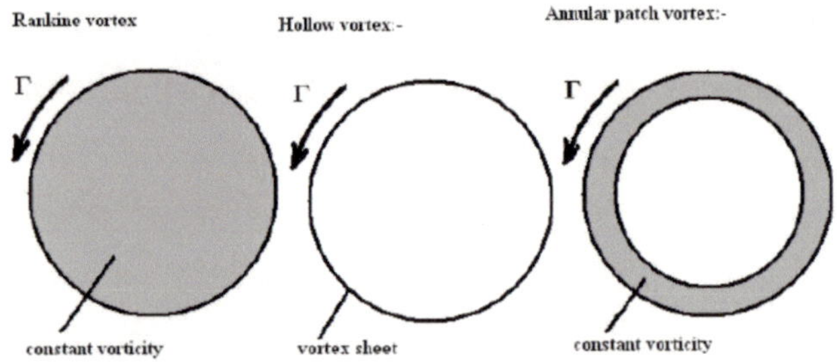


**Figure 18.**
*(left) Reference Rankine vortex core, and two annular core models, (centre) circumferential vortex sheet, (right) annular patch vortex.*

considered are shown in **Figure 18**, namely a vortex sheet (infinitely thin, located on the core circumference, consisting of a myriad of same-sign vortex elements, and hence, discrete vorticity) or an annular patch vortex, axisymmetric of finite thickness and vorticity).

The annular patch vortex velocity distribution is shown in **Figure 19**. Although vorticity would be area-proportionally greater than that of the Rankine core, there would be no effect upon $V_{MAX}$, which was the same magnitude. Inside the core edge, radial lapse rate of velocity would be much greater, as shown. The clear implication is that vorticity would need to be concentrated further, which could be only achieved by circumferentially-based discretization. This is illustrated in **Figure 20**, in three components: primary vorticity discretization (a ring of nine circumferential vortices is shown), and secondary vorticity of much lower magnitude, consisting of an annular patch vortex, to provide retrograde motion of the necklace as a whole, and an even lower constant, Rankine core vorticity, if the core is not fully diffused.

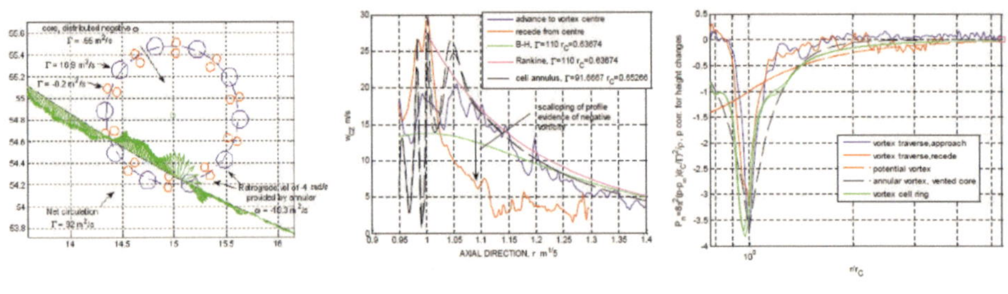

**Figure 19.**
*(left) A particular discrete necklace vortex system, matched by trial and error to the flight measured $V_V$ profile, (centre) vortex velocity; (right) vortex expansion.*

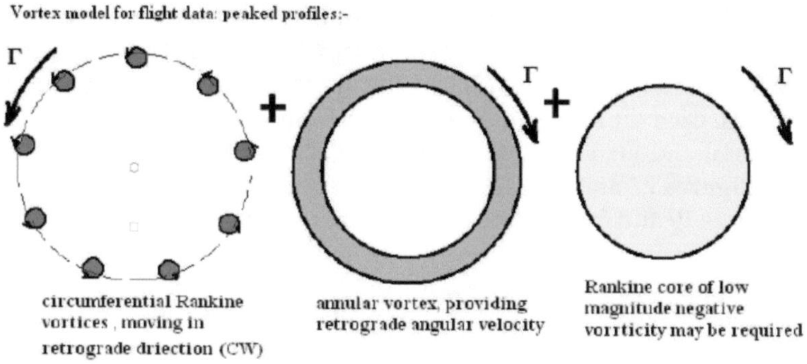


**Figure 20.**
*Example of a discrete annular vortex system for a starboard trailing vortex, primarily a necklace of discrete circumferential vortices.*

The velocity and pressure distribution of such a system, matched by trial-and-error (and hence a particular solution, rather than a general solution), are shown in **Figure 19**. $V_V$ modeling was much improved over both Rankine and BH models, at and near core edges. As observed in **Figure 19**, $r_C$ was different for the approach and recession segments, 0.83 m and 0.64 m, respectively. These $r_C$ values were greatly lower than another vortex state (**Figure 4**) on the same trailing vortex survey, namely 3.6 m. Core edge expansion modeling was likewise improved.

$V_V$ components, $V_\theta$ and $V_r$, are presented in **Figure 21**. The essential characteristics of the $V_\theta(r)$ and $V_r(r)$ flight data are the directional reversals at/inside core edges. Concerning flight data firstly (in *blue*, **Figure 21**), if a discrete circumferential vortex was in existence, $V_\theta$ must reverse direction, as $r/r_C$ is reduced <1; furthermore, $V_r$ must have maximum amplitude similar to $V_\theta$, and likewise reverse direction, in the vicinity of $r/r_C = 1$. These requirements are satisfied by the circular vortex necklace model, fitted by trial and error, located on the core edge circumference.

Therefore, a model of a ring of vortices should be demonstrable of simulating the vortex velocity flight data. The velocity magnitudes and reversals were accurately simulated, whereas the inner core flow was simulated in essential characteristics, but not accurately in detail (**Figure 21**).

At this point of the trailing vortex state identification flight data analysis, recourse to *a priori* published research was conducted. Vortex stability analysts have

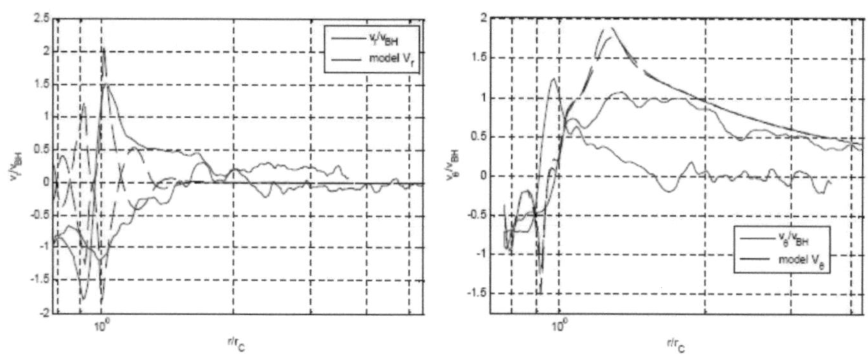


**Figure 21.**
*Vortex radial (left) and tangential (right) components, radial distributions thereof, for the vortex traverse of* ***Figure 19****.*

considered fields of a pair of co-rotating or a number of vortices as particular solutions of the Euler equations, including the dipole vortex (Lamb, [16]) and N-vortex fields (Stuart, [22]). Tur and Yanovsky [23] specifically considered N-vortex circular necklaces of vortices, each set disposed on a single circumference.

A further trial-and-error fitment of a discrete, circumferential vortex system is presented in **Figures 22** and **23**, taken from the same A388 enroute trailing vortex survey of **Figures 19** and **21**. In this case, at 3.1 m, $r_C$ was considerably greater than the first case.

As seen in **Figure 22**, the model was a stretched dipole vortex, diametrically opposite to each other, with the centre of each element of the dipole, located at $0.9r_C$.

The model over-predicted vortex velocity, $V_V$, during the advance flight-path segment into the vortex core (**Figure 23**), and accurately predicted the recession flight-path segment from the core. Recessive $V_V$ maximum was very high, 38 m/s magnitude. Average core edge expansion pressure was predicted well, greater than approach entry and less than recession exit. Inside the core, diffusion was incomplete, for both flight data and model-simulated data.

The above two examples adequately demonstrate the likely existence of particular vortex states for vented, annular vortex cores, with differing $r_C$ values, namely point-vortex ring and dipole vortex states, respectively, for two vortex core traverses.

Swaminathan et al. [24] undertook vortex merging numerical stability studies at low and high Re = Γ/υ, i.e. vortex elemental circulation, divided by kinematic viscosity.

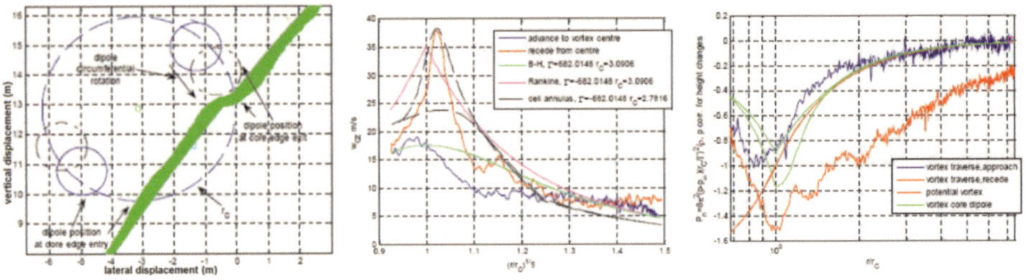


**Figure 22.**
*Additional vortex core traverse: (left) stretched dipole model; (centre) induced vortex velocity profiles; (right) core pressure expansion profiles.*

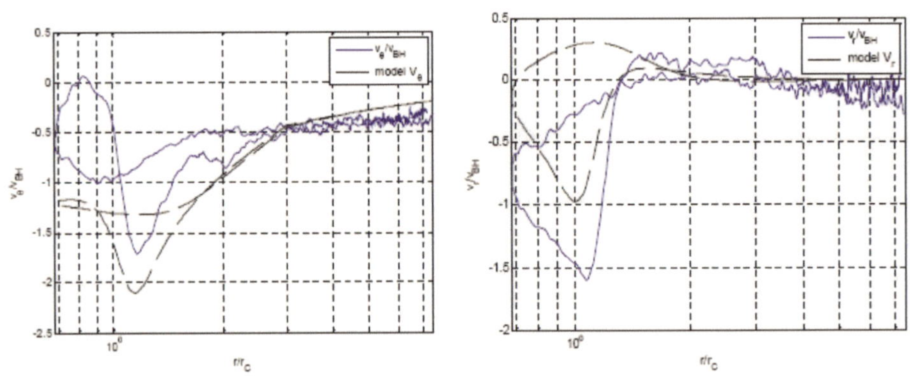


**Figure 23.**
*Dipole vortex modeling of core traverse velocity components $V_\theta$ (left) and $V_r$ (right), c.f. flight data, for the vortex core traverse of* ***Figure 22****.*

For the trailing vortices of **Figure 4** and **Figures 17–22**, $\Gamma \approx 800$ m$^2$/s and $\upsilon \approx 10^{-5}$ m$^2$/s, so that, for each vortex, Re = $8_{10}^{7}$, much higher than the [24] study cases. Nevertheless, the study may have pertinent outcomes. At high Re in the study, turbulent instabilities hastened the merging process dramatically—at Re = $2_{10}^{5}$, an N = 8 vortex ring of diameter approximately 5 m merged into a unitary vorticity annulus of diameter ≈0.8 m, in $\Delta t \approx 1$ s.

Axial flow velocities measured presently (following section) varied between a few to several m/s. Therefore, over the duration of such contracted merging, the peak expansion and hence condensate would appear of conical shape, of semi-conical angle 30–45$^0$.

**Figure 24** [14] shows the segmented, conical features of the wake vortex of a B744 transport jet. The close-up view suggested a conical ratio of $r_{min}/r_{max} \approx 1/4$, over the domain limits of the visible funnel condensate. The funnel condensate features are seen to have been aperiodically repetitive (more specifically, multi-scaled), which would in-turn require repetitive vortex reformation processes between funnels. Such reformation processes could have been ones, related to the radial instability induced outflow, under centrifugal instabilities, highlighted in Section 3.7.

The distances observed in **Figure 24**, between funnel features, would imply a rapid spatiotemporal reforming outflow process. Section 3.7 indicated high amplitude temporal rate scales in the radial instability outflow process modeled there.

Large eddy simulation (LES) modeling of extended-length, decaying trailing vortices [25] has resulted in solutions containing the appearance of funnel-shaped variations in vortex core vorticity iso-surfaces, reflective of radial/axial interaction between the outer flow and the distorted trailing vortices, prior to linking and breakup.

Although funnel condensate features were prevalent in high-altitude cruise trailing vortex condensate, thicker condensate 'blobs', which might have indicated the presence of rings of point vortices, were difficult to discern. Possibly an hexagonal pattern is discernible in the perimeter of the hollow starboard vortex core images of **Figure 25** [21].

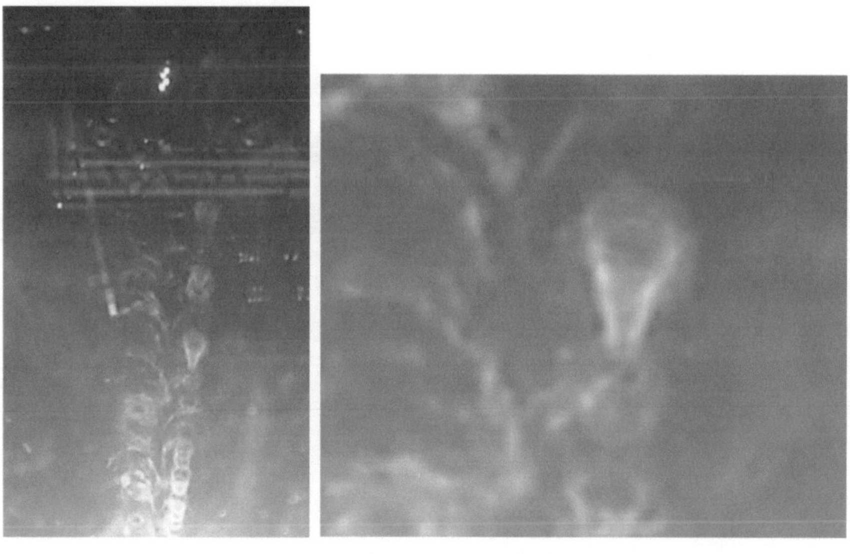

**Figure 24.**
*(left) Viewed from below, the segmented condensate of the trailing vortex pair from a B744 wake generator; (right) close-up view of a funnel feature in the starboard vortex condensate trail.*

**Figure 25.**
*Into-sun view of trailing vortices from an A388 jet transport, centred directly below the starboard vortex core.*

## 3.10 Vortex core thermodynamic structure

Associated with vortex velocity and pressure annularity, air temperature variations could also be expected. **Figure 26** presents a cross-plot of pressure and temperature variations in vortex core traverses, for an A359 survey. However, the scatter plot prevalence (**Figure 26**) was for an almost-isothermal, polytropic expansion, $Pv^n = c$, $n = 1.06$.

Rather, strong correlation was generally evidenced between heating/cooling and axial flow inside and near the core, **Figure 27**, consisting of upstream-flow heating and downstream-flow cooling.

An example of annular vortex core dynamic and thermodynamic structure is shown in **Figure 28**, for a core in the small radius state, $r_C = 1.4$ m. It consisted notably, of cooling downstream-flow in the core edge annulus, and upstream-flow heating adjacent, outside of the core.

Axial flow and temperature radial structures are shown in **Figure 29**, for a number of vented and unvented core traverses. It is seen that, for unvented cores, no particular coherence between these parameters was apparent; temperature variations were

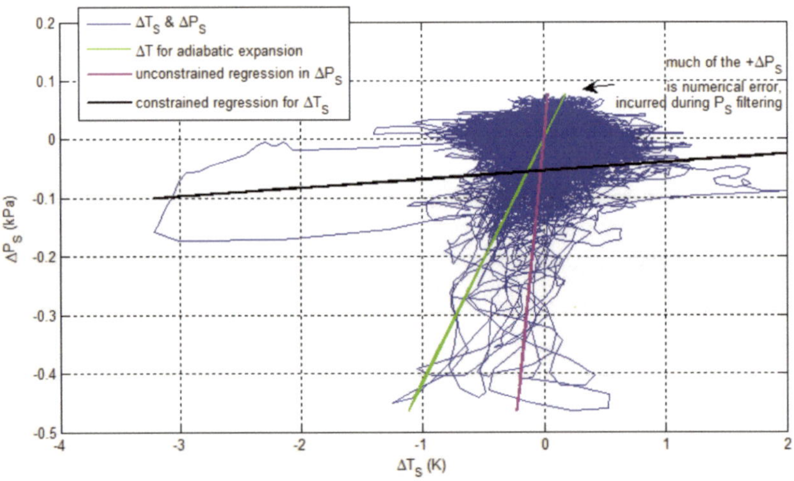


**Figure 26.**
*Temperature and pressure perturbation cross-plot, A359 vortex core traverses, showing polytropic expansion $Pv^n = c$, $n = 1.06$, with a scatter as far as adiabatic expansion, $Pv^\gamma = c$; also showing a nearly-isobaric heating and cooling mode [14].*

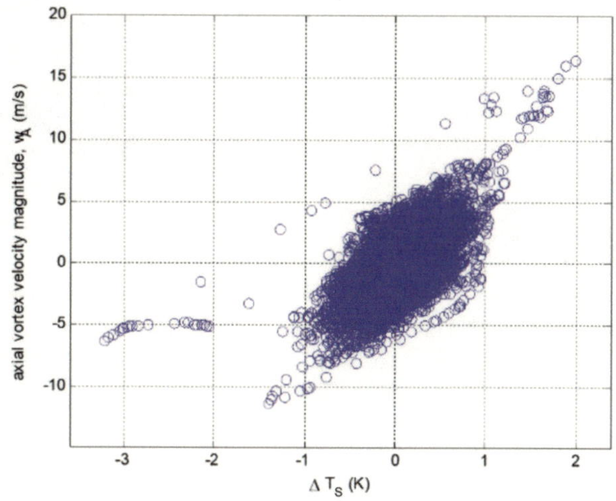


**Figure 27.**
*Correlation between axial flow and temperature, A359 vortex core traverses.*

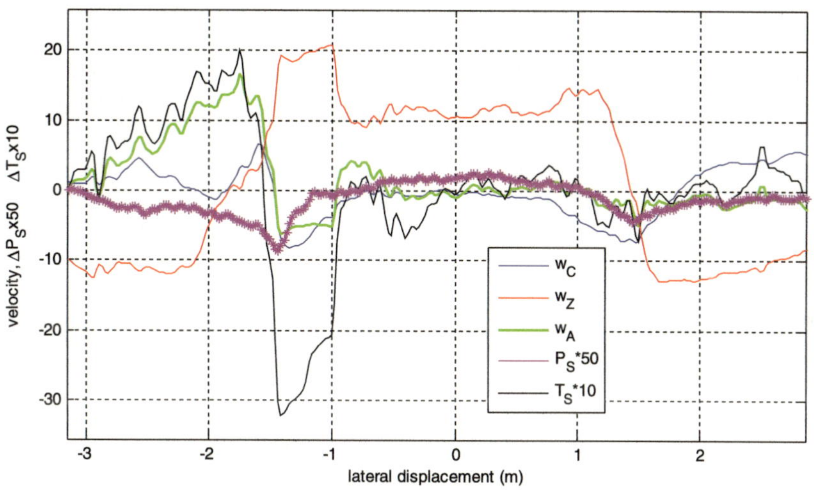


**Figure 28.**
*Vented vortex core structure—velocity, pressure and temperature structure.*

within ±0.5 K and axial velocities <±5 m/s. For vented cores, mild heating in strong upstream flow outside, adjacent to core edges and inside the core edges, strong cooling, in mild downstream flow.

## 4. Conclusions

Flight data of trailing vortices generated by heavy jet transport aircraft in cruise has been gathered by the CT-133 research aircraft of the NRC. The data has been analyzed to identify the vortex states prevalent downstream over the non-dimensional wake vortex time $t^*$ of unity to four. Lateral spacing between vortex centres was

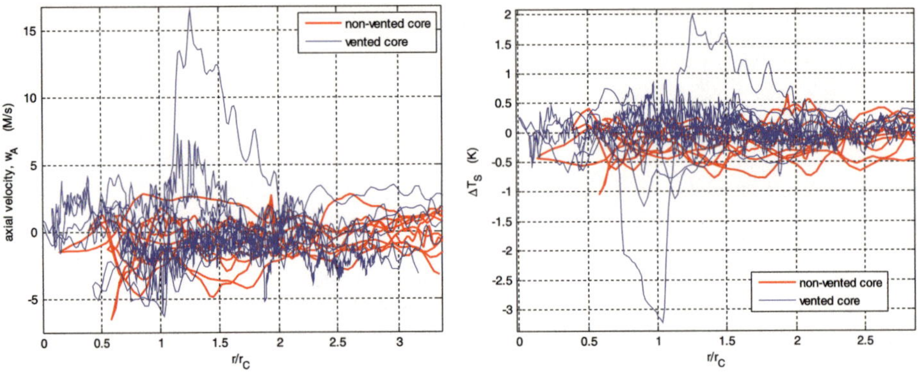


**Figure 29.**
*Vented vortex core axial flow and temperature structure, nest of traverses.*

generally 0.6–0.7 of geometric span. Vortex core radii varied greatly between several cm for small vortex elements to several metres, due to the continual change in vortex state, in short wave instability modes. There was mild growth in mean core radius to $t^* \approx 2$, followed by mild reduction.

Rounded vortex velocity radial profiles were measured, with maximum values in the Lamb-Oseen to Burnham-Hallock magnitude range of 0.8–0.5, normalized to the Rankine peak. The majority of profiles were peaked, with normalized maxima ranging 1–2. Radial instability of high magnitude was associated with the higher maxima. Numerical studies showed large values of oscillating radial flow velocities resulted from the instability, with tangential and radial flow reversals inside the core edges.

Diffused (vented) vortex cores were frequently measured—therefore evincing low vorticity inside, and annular vorticity concentrations at core edges. Peak velocities however, were higher than that from annular vorticity, indicating that it was further concentrated by discretization into N-vortex rings. Trial and error fitting to flight data profiles, was demonstrated for N = 9 and 2 (dipole). Referenced numerical studies for point-vortex rings at high Re had shown high spatiotemporal merging rates from an N = 9 ring to a single annulus of vorticity, with a ≈80% reductions in core radii. This merging has similar characteristics to funnel condensate features (up to 75% contraction in core condensate radius) in observed trailing vortices, separated by a radial outflow instability mechanism as one potential reformation process, back to larger core radii.

Vortex heating and cooling distributions were generally, nearly isobaric, whereas pressure expansion was generally, nearly isothermal. Vented cores displayed heated, upstream axial flow outside core edges, and cooled, downstream flow inside the cores.

## Acknowledgements

Support for the flight research was provided by the National Research Council Canada, and the Federal Aviation Administration of the United States.

## Author details

Anthony P. Brown
National Research Council Canada, Ottawa, Canada

*Address all correspondence to: anthony.brown@nrc-cnrc.gc.ca

## References

[1] Olsen JH, Goldberg A, Rogers M, editors. Aircraft Wake Turbulence and its Detection. London, New York: Plenum Press; 1971

[2] Caiger B, Gould DG. An analysis of flight measurements in the wake of a jet transport aircraft. In: Olsen JH, Goldberg A, Rogers M, editors. Aircraft Wake Turbulence and its Detection. London, New York: Plenum Press; 1971. pp. 125-136

[3] Brown AP, Bastian M. Enroute Wake Vortex Measurement Feasibility Flight, Operational Aspects. LTR-FR-218; 20th October. Canada, Ottawa: National Research Council; 2004

[4] Brown AP, Bastian M. Observations of wake vortex instability flowfields and aircraft response to encounters. In: AIAA-2006-6002. AIAA Atmospheric Flight Mechanics and Exhibit; 21-24 August 2006; Keystone, Colorado. DOI: 10.2514/6.2006-6002 [Published Online: 25 Jun 2012]

[5] Brown AP, Ovenden M, Bastian M. A flight study of unsteady aerodynamic loading induced by wake vortex encounter. In: AIAA AVIATION Forum; 2018 Atmospheric and Space Environment Conference; June 25-29, 2018; Atlanta, Georgia. DOI: 10.2514/6.2018-3023 [Published Online: 24 Jun 2018]

[6] Brown AP. Flight data of axial flows in wake vortex instability. In: AIAA-2016-3435. AIAA AVIATION Forum. 8th AIAA Atmospheric and Space Environments Conference; 13-17 June 2016, Washington, D.C. DOI: 10.2514/6.2016-3435 [Published Online: 10 Jun 2016]

[7] Brown AP, Bastian M. Derivation of wake vortex circulation from flight data using path integrals. In: AIAA-2007-286. 45th AIAA Aerospace Sciences Meeting and Exhibit; 8-11 January 2007; Reno, Nevada. DOI:10.2514/6.2007-286

[8] Brown AP, Crider D. Wake vortex derivation from analysis of commercial flight data. In: AIAA-2009-3867. 1st AIAA Atmospheric and Space Environments Conference; 22-25 June 2009; San Antonio, Texas. DOI: 10.2514/6.2009-3867 [Published Online: 14 Jun 2012]

[9] Brown AP. Enroute wake vortex flight data acoustic signature characteristics. In: Presentation to EU WakeNet3 Greenwake Workshop on AV and Wind Sensors; 29th-30th March 2010; Paris [Published Online: 06 Nov 2014]

[10] Mokry M. Intensification of aircraft wake vortices in cross-wind shear. AIAA Journal of Aircraft. 2003

[11] Brown AP, Mokry M, Bastian M. Flight measurements of wake vortex circulation in crossflow shear. In: AIAA-2007-285. 45th AIAA Aerospace Sciences Meeting and Exhibit; 8-11 January 2007; Reno, Nevada: American Institute of Aeronautics and Astronautics

[12] Brown AP. On the specification of wake vortex encounter gust-fields from flight data. In: AIAA-2010-7678. AIAA Atmospheric and Space Environments Conference; 2-5 August 2010; Toronto, Ontario, Canada. DOI: 10.2514/6.2010-7678 [Published Online:14 Jun 2012]

[13] Saffmann PG. Vortex Dynamics. Cambridge: Cambridge University Press; 1992

[14] Brown AP, Bastian M. WVEEFN, Wake Vortex Enroute Encounter FAA/NRC Flight Research: Wake Vortex Data Analysis. LTR-FRL-2022-0092. Flight Research Laboratory. Ottawa: NRC

[15] Brown AP. Wake Vortex Data from NASA ACCESS II. Presentation: WakeNet USA Meeting October 2015; Memphis Tennessee: Private communication

[16] Batchelor GK. An Introduction to Fluid Dynamics. Cambridge: Cambridge University Press; 1967

[17] Brown AP. Wake vortex core profiles and stability from enroute flight data. In: AIAA-2011-3032. 3rd AIAA Atmospheric Space Environments Conference; 27-30 June 2011; Honolulu, Hawaii. DOI: 10.2514/6.2011-3032 [Published Online: 14 Jun 2012]

[18] Burnham DC, Hallock JN. Chicago monostatic acoustic vortex system. Report No. DOT-TSC-FAA-79-104.IV. Cambridge MA: Transportation Systems Center; 1982

[19] Fabre D, Jacquin L. Short-wave cooperative instabilities in representative aircraft vortices. Physics of Fluids. 2004;**16**(5)

[20] Brown AP. Trailing vortex formation by lifting-line field induction and centrifugal stability analysis. In: AIAA-2012-2671. 4th AIAA Atmospheric and Space Environments; 25-29 June 2012; New Orleans, Louisiana. DOI: 10.2514/6.2012-2671 [Published Online: 06 Sep 2012]

[21] Brown AP. Vortex core structural analysis from flight data. In: Presentation: WakeNet USA, March 2013; Memphis, TN: Private communication

[22] Stuart JT. On finite amplitude oscillations in laminar mixing layers. Journal of Aircraft. 2003;**40**(2):405-407. DOI: 10.2514/2.3111

[23] Tur A, Yanovsky A. Point vortices with a national necklace: New exact stationary solutions of the two-dimensional Euler equation. Physics of Fluids. 2004;**16**(8):2877-2885

[24] Swaminathan RV et al. Dynamics of circular arrangements of vorticity in two dimensions. Physical Review E. 2016;**94**: 013105

[25] Misaka T, Holzäpfel F, et al. Vortex bursting and tracer transport of a counter-rotating vortex pair. Physics of Fluids. 2012;**24**:025104

Chapter 5

# Numerical Simulation of Energy Cascading in Turbulent Flows Using Sabra Shell Model

*Alexis Rodriguez Carranza, Obidio Rubio Mercedes and Elder Joel Varas Peréz*

## Abstract

The transfer of energy in turbulent flows occurs as a product of breaking of smaller and smaller eddies, this implies that in a spectral formulation, the transfer occurs from small wavenumbers to large wavenumbers. In order to observe the energy cascading, dissipation scales must be reached, which depend on the Reynolds number, this makes direct simulations of the Navier-Stokes equation impractical. Reduced models were investigated in recent years, such as shell models. Shell models are built by mimicking the spectral model respecting the mechanisms that are preserved, such as energy conservation, scaling and symmetries. In this paper, we will use the Sabra shell model for the study of the energy cascading in turbulent flows and we will show numerically that the energy dissipation is approximately $-1/3$ which is in agreement with the K41 theory.

**Keywords:** Navier-Stokes equations, spectral energy transfer, Kolmogorov theory, inertial range, Sabra shell model

## 1. Introduction

Kolmogorov, in 1941, formalized the idea of cascading of energy imagined by Richardson in 1922, the result of the transfer in a turbulent flow when eddies are subdivided into smaller and smaller structures until reaching very small scales where the energy is dissipated. The description of the evolution of the velocity field in a flow is given by the Navier-Stokes equations, the spectral version is given, see Ref. [1], by:

$$(\partial_t + \nu k^2)u_i(\mathbf{k}) = -\imath k_j \int \left(\delta_{i\ell} - \frac{k_i k'_\ell}{k^2}\right) u_j(\mathbf{k}') u_\ell(\mathbf{k}-\mathbf{k}')\, d\mathbf{k}' + f_i(\mathbf{k}), \tag{1}$$

where $u_i$ represents the velocity field components, $k_i$ are Fourier mode components $k$ and $\imath$ represents the imaginary unit. The problem of modeling the transfer of energy in a turbulent flow is to traverse all the scales until reaching the smallest scales, of energy dissipation, since these scales depend on the Reynolds number. To reach the dissipative scales, from Kolmogorov, $\eta$, this scale is known to be related to the

Reynolds number as $\eta \sim Re^{-3/4}$, so the mesh size $N$ to reach the scale $\eta$ grows with $Re$ what $N \sim \eta^{-3} \sim Re^{9/4}$. If we carry out a simulation in 3 dimensions, with a Reynolds number $\sim 1000$, we would have to do a mesh of the order of $10^7$. This implies that a simulation of the energy transfer for high Reynolds numbers would be unfeasible. Therefore, it is necessary to develop models that allow reducing the efforts in calculation. Lorenz in 1972 [2] was the first to give a reduced model to study 2D turbulence. The figures in this article were taken from Ref. [1]. In this article, we will use reduction models known as shell models, in which the spectral space is divided into concentric spheres, as shown in **Figure 1**.

The radii of the spheres can be considered to grow exponentially $k_n = \lambda^n$, where $\lambda > 1$ is a fixed número. The wave numbers that lie between the $n - nth$ and the sphere $(n-1) - nth$ form the $n - nth$ shell. The number of wave numbers in the $n - nth$ is of the order of $\lambda^{3n}$. In shell models, such as those proposed in Refs. [3–6], only some wave numbers with considered in each shell. In each shell, using the wave numbers that were considered in the reduction, an average velocity is obtained, which will represent the velocity in said shell. As we can see in Eq. (1), there is an interaction of three wave numbers, two on the right side and one on the left side. Shell models mimic such interaction and are constructed in such a way that inviscised invariants are conserved, by such nonlinear interaction, such as energy. The interaction coefficients couple exactly three wave numbers, the left side of (1), shows the interaction of one wave amplitude and the right side of two. A shell model from a different perspective was first proposed by Ref. [7]. The model proposed here is not obtained directly from the Navier-Stokes spectral equation, but the occurrence of an energy cascading is respected according to Kolmogorov's k41 theory. The model proposed in said work is a system of first-order ordinary differential equations with unknowns for the set of velocities, $u_n$, associated with discrete wavenumbers, $k_n$, $n = 1,2,3, \ldots$. As we mentioned before, the associated velocities of these discrete wave numbers are considered average speeds in the spectral version, $u_i(\mathbf{k})$, inside a wavenumber shell, $k_{n-1} < |\mathbf{k}| < k_n$. The equations are given by:

$$\dot{u}_n = a_{n-1}u_{n-1}u_n - a_n u_{n+1}^2 - \nu_n u_n \delta_{n>N} + f\delta_{n,1} \tag{2}$$

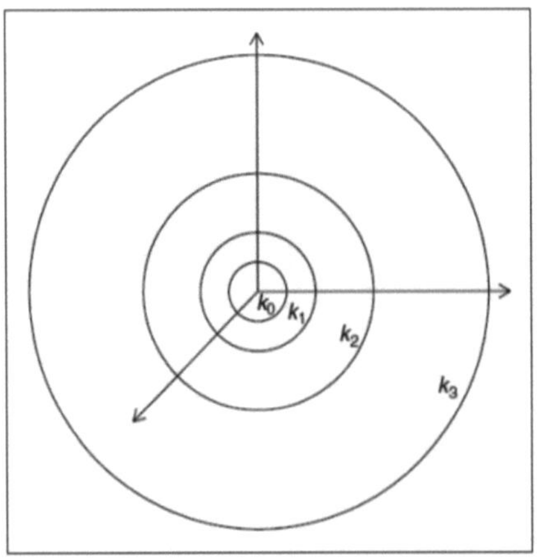


**Figure 1.**
*Division of the spectral space in concentric spheres, considering the reduction of wave numbers in the rings that are formed.*

As is known, the Navier-Stokes equation presents two mechanisms, convection and energy dissipation. The first two terms in Eq. (2) represent convection and the third term dissipation acting at large wave numbers, while the last term indicates that the force acts at small wave numbers. As in the Navier-Stokes spectral Eq. (1), the convection terms are nonlinear in the velocity, while the dissipation is linear and predominates on small scales, equivalent to large wave numbers. In a turbulent flow, energy initially acts on large scales and through the breaking of eddies it is transferred to smaller scales, showing a cascading transfer in the inertial range, to then dissipate on the Kolmogorov scales. The first two terms preserve the quantity $\sum u_n^2/2$, which represents the energy, $E$. As we mentioned, energy is an inviscid invariant, this is shown by calculating $\dot{E}$ in Eq. (2). The problem with this model is that it does not satisfy Liouville's theorem for Hamiltonian systems see Ref. [1], in the inviscid limit.

The system of Eq. (2) not satisfies Liouville's theorem trivially, this led to Ref. [3] to consider the set of equations:

$$\dot{u}_n = A_n u_{n+1} u_{n+2} + B_n u_{n-1} u_{n+1} + C_n u_{n-2} u_{n-1} - \nu_n u_n \delta_{n>N_d} + f_n, \tag{3}$$

The boundary conditions considered in said model were $u_{-1} = u_0 = 0$, $u_{N+1} = u_{N+2} = 0$.

The system (3) for coefficients $A_n, B_n, C_n$ properly chosen you get that energy, $E = \sum_n u_n^2/2$, be an inviscid invariant corresponding to turbulence 2D. Numerical simulations of the model proposed by Gledzer were made by Ref. [3]. Interest in shell models grew due to their relative simplicity for computer implementation and because they exhibited cascading energy and chaotic dynamics, characteristics that are shown from the Navier-Stokes equations. Since the shell models imitate the spectral Navier-Stokes equations, new models were proposed, the most researched being the one proposed by Yamada and Okhitani, whose current version is known as the Gledzer-Okhitani-Yamada model, GOY. It can be shown that the Navier-Stokes spectral Eq. (1), not only preserves the energy but also any quantity involving the interaction of wave numbers $\mathbf{k}, \mathbf{k}', \mathbf{k}''$, where $\mathbf{k} + \mathbf{k}' + \mathbf{k}'' = 0$.

The model used and implemented in the present work was the Sabra shell model, which is defined as a system of complex ordinary differential equations, in the velocities $u_n$, and the division of the spectral space is made in spheres of increasing radius and an average velocity is considered in each shell. The radius is considered to be increasing as $k_n = k_0 \lambda^n$, $n \geq 1$ [1, 9]. The Sabra model is given by

$$\dot{u}_n = \imath \left[ k_n u_{n+1}^* u_{n+2} - \varepsilon k_{n-1} u_{n-1}^* u_{n+1} + (1-\varepsilon) k_{n-2} u_{n-2} u_{n-1} \right] - \nu k_n^2 u_n + f_n, \tag{4}$$

where $\nu$ is the viscosity and $f_n$ is the force which is considered acting on large scales, equivalently, on small wave numbers. With boundary conditions $u_{-1} = u_0 = 0$, or $u_{N+1} = u_{N+2} = 0$.

Since we want the triads to sum to zero we can choose the wave numbers $k_n$ are chosen following the next recurrence, $k_n = k_{n-1} + k_{n-2}$, known as the Fibonacci recurrence. Using this for the wave numbers, we obtain a spacing between shells, given by $g = (\sqrt{5} + 1)/2$. Using the definition given in Ref. [8], $k_n = g^n$, being a quasi moment. The system of equations that finally given by:

$$\dot{u}_n = \imath k_n \left( u_{n+1}^* u_{n+2} - \frac{\varepsilon}{\lambda} u_{n-1}^* u_{n+1} - \frac{\varepsilon - 1}{\lambda^2} u_{n-2} u_{n-1} \right) - \nu k_n^2 u_n + f_n. \tag{5}$$

In this work we will show that the energy cascading occurs at certain scales, implementing the Sabra model, see Ref. [8] and that she obeys Kolmogorov's law.

## 2. Scale invariance of the Sabra model

The Navier-Stokes spectral equation is rescaling invariant, we can see this considering the transformation $(t, k, u(k), \nu) \to (\lambda^{1-h}t, \lambda^{-1}k, \lambda^{h+D}u(\lambda^{-1}k), \lambda^{1+h}\nu)$ for any $h$, in (1) Indeed, denoting $\overline{t} = \lambda^{1-h}t$, $\overline{k} = \lambda^{-1}k$, $\overline{u}(\overline{k}) = \lambda^{h+D}u(\lambda^{-1}k)$ and $\overline{\nu} = \lambda^{1+h}\nu$. Scaling each term of the Eq. (1) we obtain

$$\partial_t u_i(\mathbf{k}) = \partial_t \lambda^{-h-D}\overline{u}_i(\overline{\mathbf{k}}) = \lambda^{-h-D}\partial_{\overline{t}}\overline{u}_i(\overline{\mathbf{k}})\lambda^{1-h} = \lambda^{1-2h-D}\partial_{\overline{t}}\overline{u}_i(\overline{\mathbf{k}}),$$

$$k_j \int \left(\delta_{i\ell} - \frac{k_i k'_\ell}{k^2}\right) u_j(\mathbf{k}') u_\ell(\mathbf{k}-\mathbf{k}')\,\mathrm{d}\mathbf{k}'$$

$$= \lambda \overline{k}_j \int \left(\delta_{i\ell} - \lambda^2 \frac{\overline{k}_i \overline{k}'_\ell}{\lambda^2 \overline{k}^2}\right) \lambda^{-2h-2D}\overline{u}_j(\overline{\mathbf{k}}')\overline{u}_\ell(\overline{\mathbf{k}}-\overline{\mathbf{k}}')\lambda^{-D}\,\mathrm{d}\overline{\mathbf{k}}',$$

$$= \lambda^{1-2h-D}\overline{k}_j \int \left(\delta_{i\ell} - \frac{\overline{k}_i \overline{k}'_\ell}{\overline{k}^2}\right) \overline{u}_j(\overline{\mathbf{k}}')\overline{u}_\ell(\overline{\mathbf{k}}-\overline{\mathbf{k}}')\,\mathrm{d}\overline{\mathbf{k}}'$$

$$\nu k^2 u_i(\mathbf{k}) = \lambda^{-1-h}\overline{\nu}\lambda^2\overline{k}^2\lambda^{-h-D}\overline{u}_i(\overline{\mathbf{k}}) = \lambda^{1-2h-D}\overline{\nu}\overline{k}^2\overline{u}_i(\overline{\mathbf{k}}),$$

After substituting these results into Eq. (1), simplifying factor $\lambda^{1-2h-D}$ and renaming $\overline{\mathbf{k}}$ by $\mathbf{k}$, the result is followed. ■

The spectral velocities are scaled with a factor D since in the Navier-Stokes spectral equations the volume element is scaled as $d\mathbf{k} \to \lambda^{-D}d\mathbf{k}$. Therefore the shell model must mimic this scaling property. So, consider the transformation $(t, k_n, u_n) \to (\lambda^{1-h}t, k_m, \lambda^h u_m)$ as (5) and get, in the case $(\nu = f = 0)$, the following system of equations

$$\dot{u}_m = k_m a_n u_{m+1}u_{m+2} + k_m b_n u_{m-1}u_{m+1} + k_m c_n u_{m-2}u_{m-1}. \quad (6)$$

Continuing with the other symmetries of the Eq. (1) taking into account the non-viscous part, the coefficients, $n : a_n = \tilde{a}, b_n = \tilde{b}$, and $c_n = \tilde{c}$, should not be functions of the wave numbers. Taking these considerations into account, the following system of equations is obtained:

$$\dot{u}_n = k_n\left(\tilde{a}u_{n+1}u_{n+2} + \tilde{b}u_{n-1}u_{n+1} + \tilde{c}u_{n-2}u_{n-1}\right) - \nu k_n^2 u_n + f_n, \quad (7)$$

## 3. The Sabra shell model: Cascading of energy

In the Sabra model the velocities, $u_n$, are considered complex and the spectral spacing is given by $k_n = k_0\lambda^n$, see Refs. [1, 8]. Taking into account all the invariants already discussed in the previous sections, the following system of equations is proposed

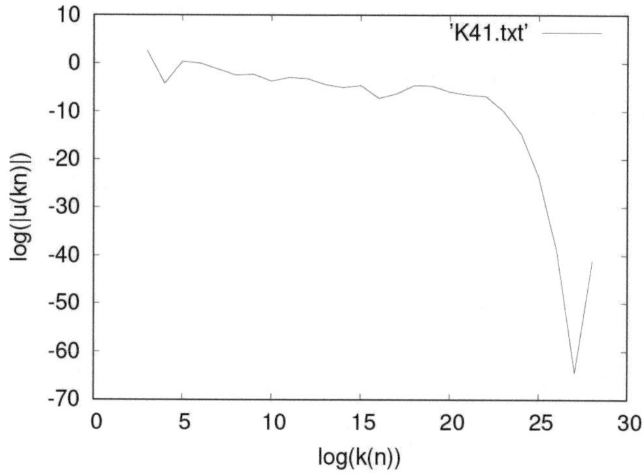


**Figure 2.**
*Shell number vs. energy on a logarithmic scale.*

$$\dot{u}_n = \imath\left[k_n u^*_{n+1} u_{n+2} - \varepsilon k_{n-1} u^*_{n-1} u_{n+1} + (1-\varepsilon) k_{n-2} u_{n-2} u_{n-1}\right] - \nu k_n^2 u_n + f_n, \qquad (8)$$

where $\nu$ is the viscosity and $f_n$ is the force which is considered acting on large scales, equivalently, on small wave numbers. With boundary conditions $u_{-1} = u_0 (= u_{N+1} = u_{N+2}) = 0$.

In the Sabra model, negative moments are defined, $k_{-n} \equiv -k_n$, and assigned the velocity, $u_{-n} = u^*_n$, at these moments. Since the velocities are complex, from Eq. (8) that the number of unknowns is 2$N$, $N$ for the real part and $N$ for the complex part of the velocities. Así (8) can be written as

$$\left(\mathrm{d}/\mathrm{d}t + \nu k_n^2\right) u_n = \imath k_n \sum_{k_\ell < k_m} I(\ell, m; n) u_\ell u_m + f_n, \qquad (9)$$

where, see Ref. [1],

$$I(\ell, m; n) = \delta_{n+1,\ell}\delta_{n+2,m} - \frac{\varepsilon}{\lambda}\delta_{n-1,\ell}\delta_{n+1,m} + \frac{1-\varepsilon}{\lambda}\delta_{n-2,\ell}\delta_{n-1,m}.$$

### 3.1 Numerical simulation

For the numerical simulation, they used the boundary conditions $u_{-1} = u_0 = 0$, or $u_{N+1} = u_{N+2} = 0$. The results of wavenumbers vs. energy are plotted in **Figure 2**. Below is a part of the code.

Listing 1.1. Fortran code.

```
program sabra
!Declaration of variables
!Initializing RK4 for future velocities
do j = 1,n
    do i = 1,m
        call du(j,n,m,x,F)
            K1(k) = h*F(k)
```

```
                K2(k) = h*F(k)
                K3(k) = h*F(k)
                K4(k) = h*F(k)
        end do

        w(j + 1,i) = x(j,i) + (1./6.)*(K1(i) + 2*K2(i) + 2*K3(i) + K4(i))
        x(j + 1,i) = w(j + 1,i)
    end do

    subroutine du(j,n,m,x,F)
     integer, intent(in)::j,n,m
     real, intent(in):: x(n + 1,m)
     real, intent(out)::F(m)
       F(1) = -4.*x(j,1) + 3.*x(j,2) + 6.
        F(2) = -2.4*x(j,1) + 1.6*x(j,2) + 3.6
end subroutine
```

We observe in **Figure 2** that energy is transferred with a linear angular coefficient in certain shells, and scales in physical space, as prescribed in Kolmogorov's theory. Doing a fit or linear regression on these scales, we obtain an angular coefficient of $-1.36$. Highlighting that the K41 theory [9] predicts an angular coefficient of $-1.66\ldots$. Said fit is obtained in **Figure 3**.

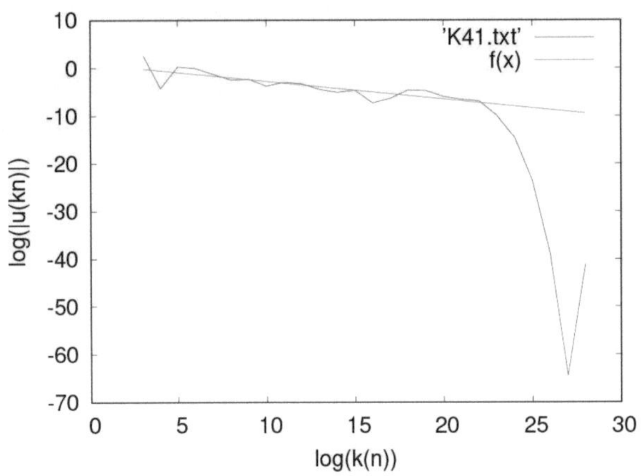


**Figure 3.**
*Fit or linear regression where the energy transfer occurs, obtaining an angular coefficient of* $-1.36$.

## 4. Conclusions

- The energy cascading occurs at certain scales, inertial range, which is visible in our simulation.
- The angular coefficient obtained was $-1.36$ which is approximated the angular coefficient prescribed by the theory $K41$ of Kolmogorov, $-1.6$.
- In turbulent fluids, the presence of an energy cascading was corroborated.

## Author details

Alexis Rodriguez Carranza[1*], Obidio Rubio Mercedes[1] and Elder Joel Varas Peréz[2]

1 National University of Tujillo, Trujillo, Perú

2 National University Jorge Basadre-Tacna, Tacna, Perú

*Address all correspondence to: arodriguezca@unitru.edu.pe

## References

[1] Ditlevsen PD. Turbulence and Shell Models. Cambridge University Press; 2010

[2] Lorenz EN. Low order models representing realizations of turbulence. Journal of Fluid Mechanics. 1972;**55**: 545-563

[3] Yamada M, Ohkitani K. Lyapunov spectrum of a chaotic model of three-dimensional turbulence. Journal of the Physical Society of Japan. 1987;**56**: 4210-4213

[4] Jensen MH, Paladin G, Vulpiani A. Intermittency in a cascade model for three-dimensional turbulence. Physical Review A. 1991;**43**:798

[5] Pissarenko D, Biferale L, Courvoisier D, Frisch U, Vergassola M. Further results on multifractality in shell models. Physics of Fluids A Fluid Dynamics. 1993;**5**:2533

[6] Benzi R, Biferale L, Parisi G. On intermittency in a cascade model for turbulence. Physica D. 1993;**65**:163-171

[7] Obukhov AM. On some general characteristics of the equations of the dynamics of the atmosphere. Izv. Akad. Nauk SSSR, Fiz. Atmos. Okeana. 1971;**7**: 695-704

[8] L'vov VS, Podivilov E, Pomyalov A, Procaccia I, Vandembroucq D. Improved shell model of turbulence. Physical Review E. 1998;**58**:1811-1822

[9] Kolmogorov AN. The local structure of turbulence in incompressible viscous fluid for very large Reynolds number. Doklady Akademii Nauk SSSR. 1941;**30**: 301-305

Chapter 6

# Algorithm to Generate Liutex Core Lines Based on Forward Liutex Magnitude Gradient Lines

*Yifei Yu and Chaoqun Liu*


## Abstract

Vortex definition and identification are extremely important for the study of fluid dynamics research. Liutex is a newly proposed concept that correctly represents vortex. Liutex is a vector whose direction is the local rotation axis and whose magnitude is twice the angular speed. To identify the unique structure of a vortex, a method known as the Liutex Core Line method has been developed, which displays the rotational core axis of a vortex. However, the original method is a manual method, which is not practical for real application, and an automatic algorithm is required for practical usage. Xu et al. proposed an algorithm by selecting the best line from a group of candidate lines, which is an important progress. In this chapter, from another perspective to solve this problem, a new algorithm is introduced based on forward Liutex magnitude gradient lines. Since gradient lines have the feature that they advance to the local maximums, the route will still result in a unique line, which avoids the process to find the best line. This algorithm has achieved some success for the Lambda vortex in early boundary layer transition.


**Keywords:** Liutex, vortex, vortex core, vortex identification, Liutex lines, Liutex core line

## 1. Introduction

Defining and visualizing vortices have been a challenge in fluid mechanics for several decades, even though the vortex is the essential part of formulating the turbulent flow that consists of countless vortices in different sizes and strengths. Turbulence generation and sustenance are still a mystery after intensive research efforts of centuries. Kaczorowski et al. [1] and Xi et al. [2] performed deep theory and experiment research on this topic, especially the Rayleigh–Bénard (RB) convection. In 2019, Zhou et al. elaborated the hydrodynamic instabilities that induced turbulent mixing in wide areas including inertial confinement fusion, supernovae, and their transition criteria. Zhou [3, 4] described in detail Rayleigh–Taylor and Richtmyer–Meshkov instability and some related models. Zhou's analysis is systematical, comprehensive, and sophisticated for flow instability and vortex generation, covering long history and state-of-the-art advances in turbulence research, which has clearly shown guidance


IntechOpen

for further and deeper scientific research. Since vortex is the sinus and muscles of turbulence, it is of great importance to find out the definition of vortex. Human's understanding of vortex has gone through three stages [5].

The first generation is vorticity-based methods which have several weaknesses. One severe weakness is the contamination by shear. This shear contamination can be easily seen in the near-wall laminar boundary layer where there are high levels of shear with small amounts of rotational motion. High vorticity magnitude is seen in areas of high shear and low rotation, while areas with stronger rotation can be found to have relatively lower vorticity magnitude values [6]. Though the misconception that a vortex is defined by vorticity is still being published in textbooks and research papers, we know this is not appropriate. Then, the second generation that uses eigenvalue-based vortex identification criteria was developed. The second generation of vortex identification methods mainly used the eigenvalues of the velocity gradient tensor. Among the second generation, there are some popular methods such as Q criterion [7], $\Delta$ criterion [8], $\lambda_{ci}$ criterion [9], $\lambda_2$ criterion [10], and some other methods. Although these methods made advances over the first generation, they were still contaminated by shear in different degrees. Another problem is that they are all scalar-based. These methods require a user-specified threshold to draw an iso-surface. Since the threshold was somewhat arbitrary, there must be many different thresholds used to represent the vortical structure and different thresholds may give totally different vortical structures. These scalar-based methods also prevent us from determining the rotation axis except that in the paper of $\lambda_{ci}$ [9], Zhou et al. indicated the rotation axis direction aligns the eigenvector but $\lambda_{ci}$ was still defined as a scalar and it is contaminated by shear as well due to the fact that the non-orthogonal transformation is used in the definition. To find the proper definition of vortex, Liu et al. proposed Liutex [11], which is a vector, in 2018. The direction of Liutex is the local rotation axis, and the magnitude of Liutex represents twice the angular speed.

Since the invention of Liutex, many researchers have used and tested it and reported that it is a strictly mathematical vortex definition and the currently best vortex identification method. Cuissa et al. [12] found that Liutex matches the analytical result of Lamb-Osceen vortex. Shen et al. [13] commented that Liutex is the only method that fully identifies small-scale vortices in vertical slit fishways simulation. Xu et al. found Liutex similarity, i.e., both the frequency and wavenumber spectrum of Liutex match the -5/3 law [14–16]. More reports on Liutex application can be found from [17–29] and four published books [30–33]. Liutex is considered as the vortex definition in this chapter.

Since Liutex is a vector-based concept that provides the direction as well as the magnitude of the rigid rotation, we can integrate the Liutex vector passing through a point within the vortex region to obtain a special Liutex core line. The concept of Liutex core line was first proposed in [34]. In short, Liutex core lines are formed by points where local maximal Liutex magnitude is reached. When calculating Liutex core lines using computer programs, some numerical difficulties are raised. Because of the numerical errors, the points selected by Liutex core line definition are a cluster of points rather than a single point. To solve this problem, Xu et al. [35] proposed seed point selection method. Li et al. [36] further improved this method by choosing the best line among the candidate lines generated by seed points. In this chapter, a new method, from another perspective, is reported to calculate the Liutex core lines based on the idea that the forward Liutex magnitude gradient lines converge to the Liutex core lines. So, instead of drawing two-directional Liutex magnitude lines, we only draw the forward lines. This leads to one advantage that the converging lines are unique so that people do not need to select one line from several candidate lines. This method achieves preliminary success for identifying Liutex core lines of the Lambda vortex in early boundary layer transition.

This chapter is organized as follows: The definitions of Liutex and Liutex core lines are introduced in Section 2 and 3, respectively. In Section 4, the manual method and Xu's algorithm are reviewed. Then, the new proposed algorithm is introduced in Section 5. In Section 6, the proposed automatic method is applied to find the Liutex core lines from the data obtained from a Direct Numerical Simulation (DNSUTA) [37] of flow transition in a boundary layer validated by researchers from UTA and NASA Langley. The results of the DNSUTA were also compared to other DNS results where the DNSUTA was shown to have remarkable consistency with [38]. Conclusions are made in Section 7.

## 2. Liutex

Liutex $\vec{R}$ is a vector such that

$$\vec{R} = R\vec{r} \tag{1}$$

where its direction $\vec{r}$ is the eigenvector of the velocity gradient tensor $grad\vec{v}$ that satisfies $\vec{\omega} \cdot \vec{r} > 0$ where $\vec{\omega}$ is the vorticity and the magnitude R can be calculated by the following equation:

$$R = \left(\vec{\omega} \cdot \vec{r}\right) - \sqrt{\left(\vec{\omega} \cdot \vec{r}\right)^2 - 4{\lambda_{ci}}^2} \tag{2}$$

$\vec{r}$ is the real eigenvector of the velocity gradient tensor matrix representing local rotation axis, and $R$ refers to the twice angular speed.

## 3. Liutex core line

The Liutex core line [34] (or the vortex rotation axis) can be defined as a Liutex line passing through the points that satisfy the condition:

$$\nabla R \times \vec{r} = 0, R > 0 \tag{3}$$

where $\nabla R$ represents the Liutex magnitude gradient vector and $\vec{r}$ represents the direction of the Liutex vector. The Liutex core lines can be colored according to the vortex strength (Liutex magnitude) and provide a unique representation of the vortex structure.

The Liutex core concept comes from the idea that Liutex core is located at the position where Liutex reaches the maximum in the plane perpendicular to the local Liutex direction as shown in **Figure 1**. Since it is the local maxima in the plane, the projection of the Liutex gradient in the plane is zero, and thus, its only possible direction is perpendicular to the plane, which is parallel to the local Liutex direction. So, Eq. (3) is used to define Liutex core.

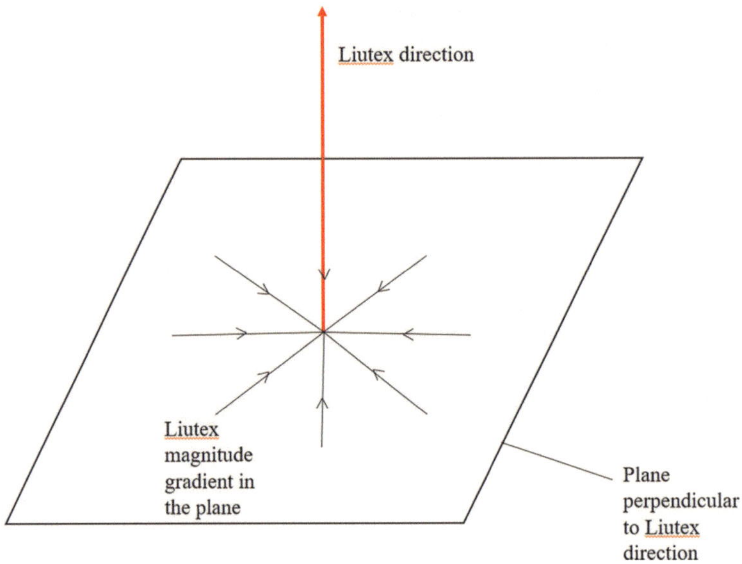


**Figure 1.**
*Liutex magnitude gradient lines distribution for the local maximal Liutex magnitude in the plane perpendicular to the Liutex direction.*

## 4. Manual method and Xu's algorithm to calculate Liutex core lines

After defining the Liutex core line in [34], a manual method was proposed simultaneously in the same paper based on visualization software, for instance, Tecplot, Paraview, and so forth. After loading data files into these software, slices, iso-surfaces, and streamlines can be drawn, which are used in the manual method. The steps of the manual method are listed below.

Manual method:

Step 1: Draw the iso-surface of Liutex to get a general idea of the vortex structure.

Step 2: Set a slice that exhibits the contour of Liutex magnitude.

Step 3: Determine the concentration of line of some Liutex magnitude gradient lines passing through the slice.

Step 4: Find the intersection point of the concentration line and the slice and create a Liutex line passing through the intersection point.

Use Burgers vortex as an example. The Bugers vortex is an analytical solution of the Navier-Stokes governing equations and is a typical widely used vortex model. Its velocity field can be described in the cylindrical coordinates as follows:

$$v_r = -\alpha r \tag{4}$$

$$v_z = 2\alpha z \tag{5}$$

$$v_\theta = \frac{\Gamma}{2\pi r} g(r) \tag{6}$$

where $\alpha > 0$ and $\Gamma > 0$ are constants, and $g(r)$ can be expressed as

$$g(r) = 1 - e^{\left(-\frac{\alpha r^2}{2\nu}\right)} \tag{7}$$

Some streamlines of the Burgers vortex is shown in **Figure 2**. It can be seen that Burgers vortex is like an upside-down cyclone where the flow converges into the center and then stays in the center and is stretched. As stated in the step 1, the iso-surface of Liutex is shown in **Figure 3**, which shows there is a vortex center inside the iso-surface. Then, according to step 2 and step 3, a slice cutting the iso-surface is created, and gradient of Liutex magnitude lines is drawn, as shown in **Figure 4**. Clearly, all gradient lines converge to one point, and that point is the seed point. Next, we draw the Liutex line passing through the seed point, and this line represents the vortex center, and the result is shown in **Figure 5**.

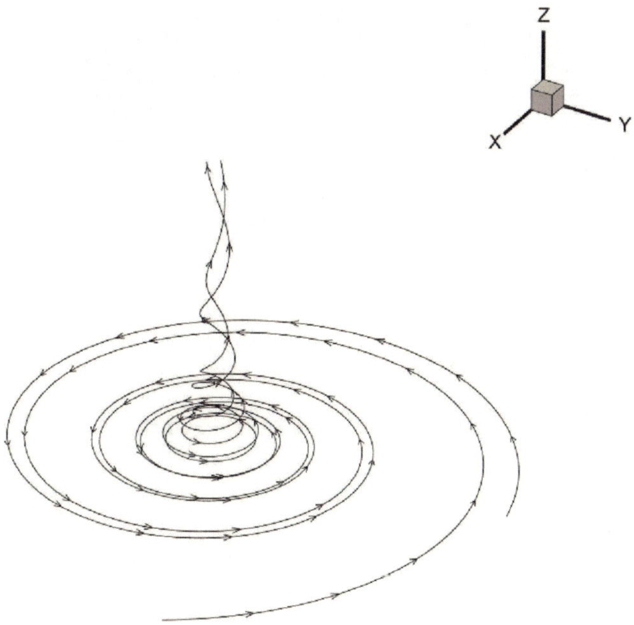


**Figure 2.**
*Streamlines of the Burgers vortex.*

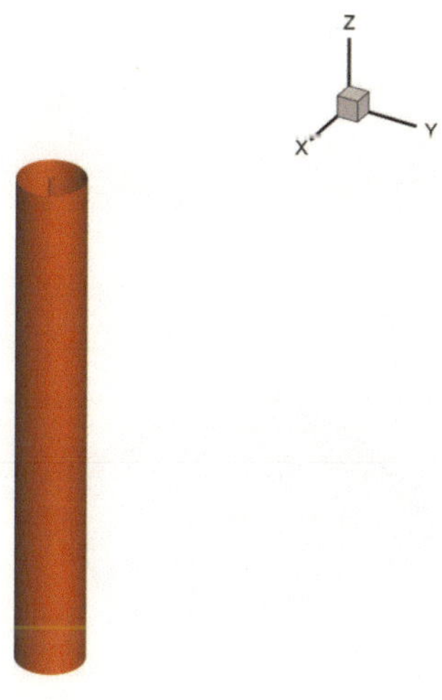


**Figure 3.**
*Liutex iso-surface of the Burgers vortex.*

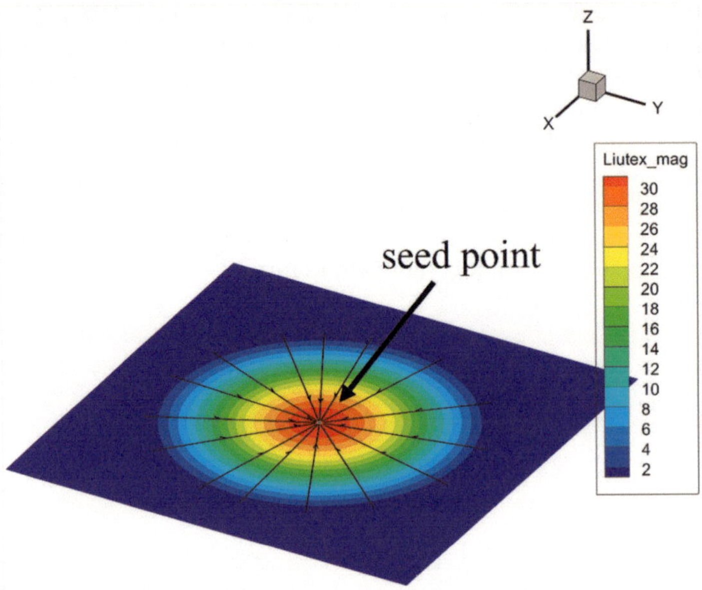


**Figure 4.**
*Liutex magnitude distribution on a slice with Liutex gradient lines.*

The Burgers vortex is a simple vortex model. Then, a more complicated case is tested. We will use the direct numerical simulation of the flat plate boundary layer transition. Similarly, we can first draw the iso-surface of Liutex (shown in **Figure 6**) and find the vortex structures are much more complicated than the Burgers vortex. Next, we display Liutex gradient lines and find the intersection point of the lines and the slice as shown in **Figure 7**. Then, we draw the Liutex line passing through that point. In **Figure 8**, colors are used to represent the Liutex magnitude and we can see the rotation strengths.

Obviously, the manual method is low in efficiency and impractical for generating all Liutex core lines for complicated vortex. So, many scientists aim to find an automatic way. Li and Xu et al. proposed an algorithm to automatically calculate Liutex core lines in [36].

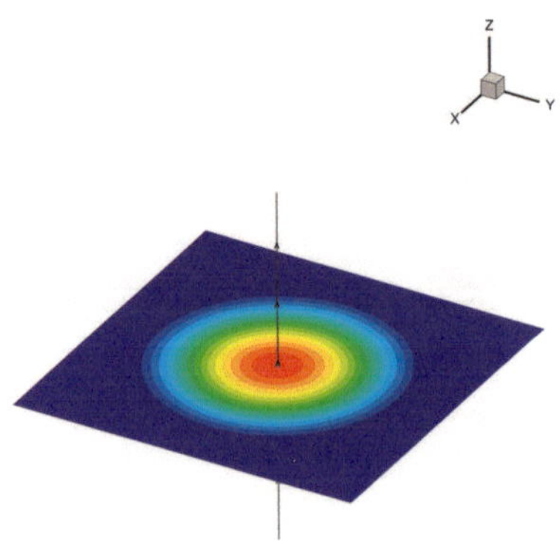


**Figure 5.**
*Liutex core line of the Burgers vortex.*

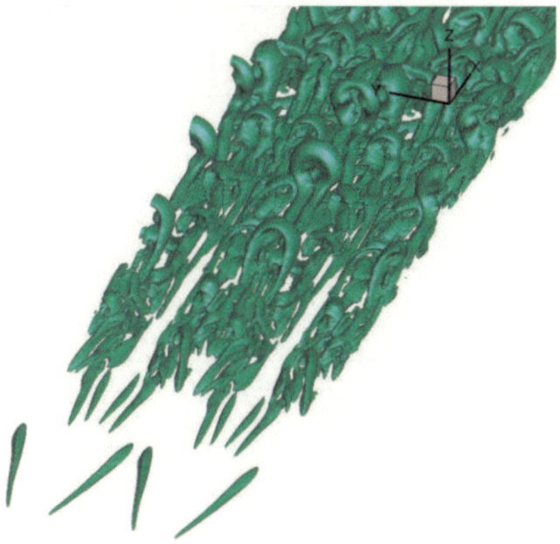

**Figure 6.**
*Liutex iso-surface of the flat plate boundary layer transition.*

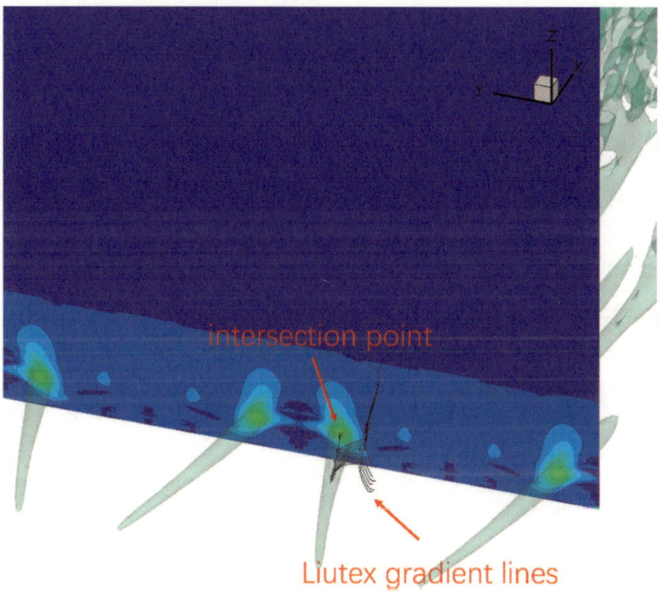


**Figure 7.**
*Liutex iso-surface, a slice, Liutex gradient lines, and the intersection point.*

Xu's algorithm:

Step 1: Read position and velocity data.

Step 2: Compute Liutex vector.

Step 3: Search vortex core points by the criterions $\left|\nabla R \times \vec{r}\right| < \varepsilon$ and $\Omega_R > \Omega_0$ where $\Omega_R$ represents relative rotation strength and can be calculated by

$$\Omega_R = \frac{\left(\vec{\omega} \cdot \vec{r}\right)^2}{2\left[\left(\vec{\omega} \cdot \vec{r}\right)^2 - 2\lambda_{ci}^2 + 2\lambda_{cr}^2 + \lambda_r^2\right] + \varepsilon} \tag{8}$$

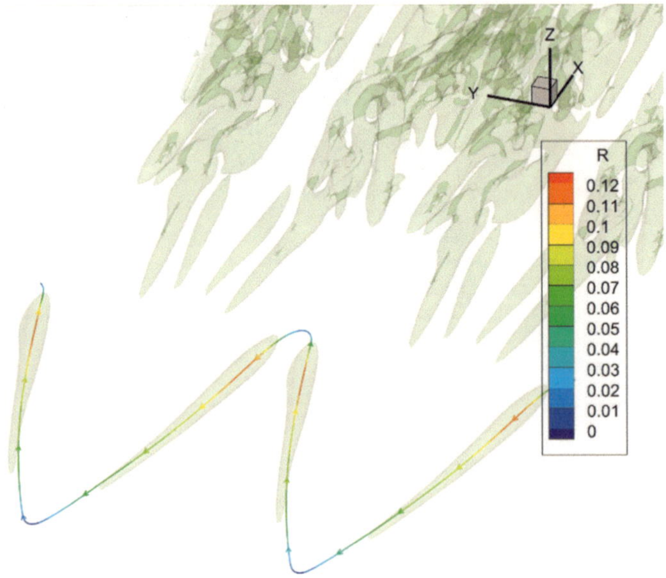


**Figure 8.**
*Liutex iso-surface and Liutex core line.*

where $\varepsilon$ is a small positive number, $\lambda_{cr}$ and $\lambda_{ci}$ are the real and complex eigenvalues of the velocity gradient tensor, and $\lambda_r$ is the real eigenvalue of the velocity gradient tensor.

Step 4: Compute vortex core lines based on selected points in step 3.

Step 5: Group the vortex core lines.

Step 6: Delete the false and inaccurate vortex core lines by the third criterion.

Step 7: Output data of vortex core lines.

To make it simple, Xu's algorithm indeed proposes a series of criterions on how to select the seed points. His method first selects some seed point candidates by applying some criterions. And then, it groups the seed point candidates of the same vortex together. After that, it selects the best one from the seed point candidates in each group.

## 5. New algorithm to calculate Liutex core line

As explained in Section 3, Liutex direction is parallel to Liutex magnitude gradient direction at vortex core positions, so theoretically, Liutex lines and Liutex gradient lines will be the same in the location where the Liutex core lines pass. However, in practice, due to numerical errors, the points that the numerical program finds will not be exactly located on the same single Liutex line, which requires selecting the best one by a specified algorithm in Xu's paper. Another perspective to solve this problem is to make use of the property of gradient lines. Gradient lines have a good property that they can converge to the local maximums. In **Figure 9**, people can see that even though the seed points are much away from the vortex center, the generated Liutex gradient lines will converge to a unique line finally. So, even though selected points have numerical errors, we will still get a unique line. Based on this idea, a new algorithm is proposed as follows.

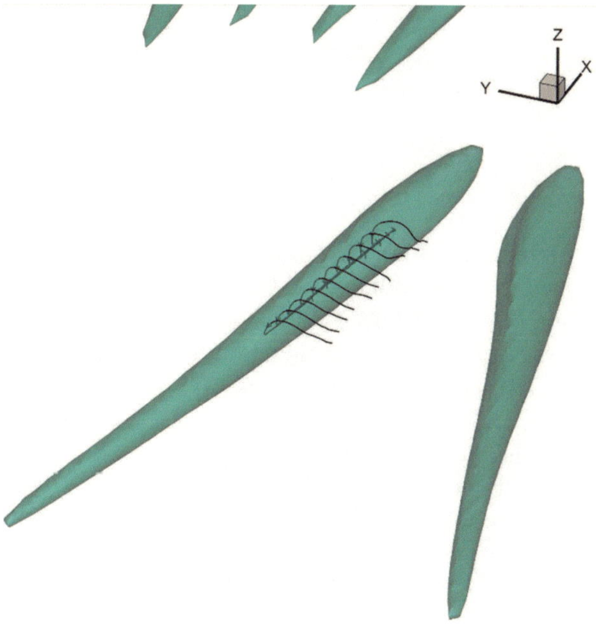


**Figure 9.**
*Liutex iso-surface and Liutex gradient lines.*

Step 1: Set the Liutex threshold and exclude the non-vortex points whose Liutex magnitudes are below the threshold.

Step 2: Find all seed points satisfying

$$\left|\nabla R \times \vec{r}\right| < \varepsilon \tag{9}$$

where $\varepsilon$ is a small number. Originally, the Liutex core line definition is $\nabla R \times \vec{r} = 0$. Because computer has numerical errors, we set a small number $\varepsilon$ rather than 0.

Step 3: Draw forward Liutex gradient lines starting from the selected seed points.

The proposed method has the following merits. First, this method does not have a high demand on the accuracy of the seed points. Xu's method already shows that it is hard to get enough accurate seed points and to ensure a unique line, people need to pick one seed point from seed point candidates. But the proposed method can tolerate some errors on the seed point selection. Secondly, since the proposed method can tolerate some errors on the seed point selection, the parameters in this method are not case-sensitive because people do not rely on the criterions to filter out all fake seed points.

## 6. Test case

In this section, the proposed automatic method is applied to find the Liutex core line for the data obtained from a Direct Numerical Simulation [6] (DNSUTA) of flow transition in a boundary layer, which has been validated by researchers from UTA and

| $M_\infty$ | Re | $x_{\text{in}}$ | $Lx$ | $Ly$ | $Lz_{\text{in}}$ | $T_w$ | $T_\infty$ |
|---|---|---|---|---|---|---|---|
| 0.5 | 1000 | $300.79\delta_{in}$ | $798.03\delta_{in}$ | $22\delta_{in}$ | $40\delta_{in}$ | 273.15 K | 273.15 K |

**Table 1.**
*Flow parameters.*

NASA Langley. The mesh of this simulation includes 1920 × 128 × 241 points in streamwise (x), spanwise (y), and wall normal (z) directions. The parameters of the flow are shown in **Table 1**. $M_\infty$ is the inflow Mach number; Re represents Reynolds number, which is defined as $Re = \frac{\rho_\infty u_\infty \delta_{in}}{\mu_\infty}$ using the inflow boundary layer displacement thickness as the length reference; $x_{in}$ is the distance between the leading edge of the flat plate and upstream boundary of the computational domain; $Lx$ and $Ly$ are the lengths of the computational domain along $x$ and $y$ directions, respectively; $Lz_{in}$ means the height at inflow boundary; $T_w$ and $T_\infty$ represent the temperatures of wall and free stream, respectively. The illustration of the computational domain is shown in **Figure 10**.

This simulation uses the 3D compressible Navier–Stokes equations in curvilinear coordinates, which can be expressed as the following,

$$\frac{1}{J}\frac{\partial \mathbf{Q}}{\partial t} + \frac{\partial(\mathbf{E}-\mathbf{E}_v)}{\partial \xi} + \frac{\partial(\mathbf{F}-\mathbf{F}_v)}{\partial \eta} + \frac{\partial(\mathbf{H}-\mathbf{H}_v)}{\partial \zeta} = 0 \tag{10}$$

$$\mathbf{Q} = \begin{pmatrix} \rho \\ \rho u \\ \rho v \\ \rho w \\ e \end{pmatrix}, \boldsymbol{E} = \frac{1}{J}\begin{pmatrix} \rho U \\ \rho u U + p\xi_x \\ \rho v U + p\xi_y \\ \rho w U + p\xi_z \\ (e+p)U \end{pmatrix}, \boldsymbol{F} = \frac{1}{J}\begin{pmatrix} \rho V \\ \rho u V + p\eta_x \\ \rho v V + p\eta_y \\ \rho w V + p\eta_z \\ (e+p)V \end{pmatrix} \tag{11}$$

$$\boldsymbol{H} = \frac{1}{J}\begin{pmatrix} \rho W \\ \rho u W + p\zeta_x \\ \rho v W + p\zeta_y \\ \rho w W + p\zeta_z \\ (e+p)W \end{pmatrix}, \boldsymbol{E}_v = \frac{1}{J}\begin{pmatrix} 0 \\ \tau_{xx}\xi_x + \tau_{yx}\xi_y + \tau_{zx}\xi_z \\ \tau_{xy}\xi_x + \tau_{yy}\xi_y + \tau_{zy}\xi_z \\ \tau_{xz}\xi_x + \tau_{yz}\xi_y + \tau_{zz}\xi_z \\ q_x\xi_x + q_y\xi_y + q_z\xi_z \end{pmatrix} \tag{12}$$

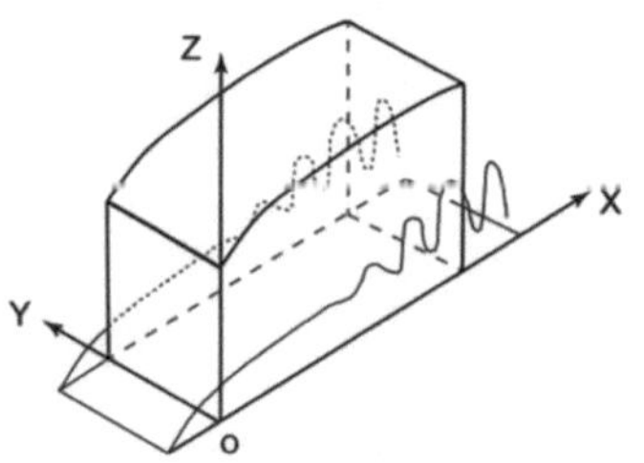


**Figure 10.**
*Computational domain.*

$$\boldsymbol{F}_v = \frac{1}{J}\begin{pmatrix} 0 \\ \tau_{xx}\eta_x + \tau_{yx}\eta_y + \tau_{zx}\eta_z \\ \tau_{xy}\eta_x + \tau_{yy}\eta_y + \tau_{zy}\eta_z \\ \tau_{xz}\eta_x + \tau_{yz}\eta_y + \tau_{zz}\eta_z \\ q_x\eta_x + q_y\eta_y + q_z\eta_z \end{pmatrix}, \quad \boldsymbol{E}_v = \frac{1}{J}\begin{pmatrix} 0 \\ \tau_{xx}\zeta_x + \tau_{yx}\zeta_y + \tau_{zx}\zeta_z \\ \tau_{xy}\zeta_x + \tau_{yy}\zeta_y + \tau_{zy}\zeta_z \\ \tau_{xz}\zeta_x + \tau_{yz}\zeta_y + \tau_{zz}\zeta_z \\ q_x\zeta_x + q_y\zeta_y + q_z\zeta_z \end{pmatrix} \tag{13}$$

where $J$ is the Jacobian of the coordinate transformation; $\xi_x, \xi_y, \xi_z, \eta_x, \eta_y, \eta_z, \zeta_x, \zeta_y, and\ \zeta_z$ are coordinate transformation metrics; $\boldsymbol{Q}$ is the vector of conserved quantities; $\boldsymbol{E}, \boldsymbol{F}, and\ \boldsymbol{H}$ are the inviscid flux vectors and $\boldsymbol{E_v}, \boldsymbol{F_v}, and\ \boldsymbol{H_v}$ are the viscous vectors; and $\tau_{xx}, \tau_{yx}, \tau_{zx}, \tau_{xy}, \tau_{yy}, \tau_{zy}, \tau_{xz}, \tau_{yz}, and\ \tau_{zz}$ are viscous stress components.

Compact schemes are used to do the spatial discretization. It has the following form:

$$\beta_- f'_{j-2} + \alpha_- f'_{j-1} + f'_j + \alpha_+ f'_{j+1} + \beta_+ f'_{j+2} = \frac{1}{h}\left(b_- f_{j-2} + a_- f_{j-1} + c f_j + a_+ f_{j+1} + b_+ f_{j+2}\right) \tag{14}$$

where $f'_j$ is the derivative at point j. For the sixth-order scheme used in this simulation,

$$\beta_- = 0, \alpha_- = \frac{1}{3}, \beta_+ = 0, \alpha_+ = \frac{1}{3}, b_- = -\frac{1}{36}, a_- = -\frac{7}{9}, a_+ = \frac{7}{9}, b_+ = \frac{1}{36} \tag{15}$$

The total variation diminishing (TVD) third order Runge–Kuntta method is used for the time discretization. The equations are:

$$Q^{(0)} = Q^{(n)} \tag{16}$$

$$Q^{(1)} = Q^{(0)} + \Delta t R^{(0)} \tag{17}$$

$$Q^{(2)} = \frac{3}{4}Q^{(0)} + \frac{1}{4}Q^{(1)} + \frac{1}{4}\Delta t R^{(1)} \tag{18}$$

$$Q^{(n+1)} = \frac{1}{3}Q^{(0)} + \frac{2}{3}Q^{(2)} + \frac{2}{3}\Delta t R^2 \tag{19}$$

The simulation result is shown in **Figure 11**. Spanwise vortex appears first followed by the Lambda vortex. We will use the Lambda vortex region to test the proposed method.

Applying step 1 and step 2 of the proposed method, we can get the seed points as shown in **Figures 12–14**. Drawing forward Lituex gradient lines starting from these seed points, we get **Figures 15–17**. If we draw Liutex lines from these seed points, we will get several separate lines rather than a unique line as shown in **Figures 18–20**. The proposed new algorithm obtains the unique vortex core line and does not need to select the best line from a group of lines generated by the seed points. Even though the obtained seed points have numerical errors, the output will still be convergent to a

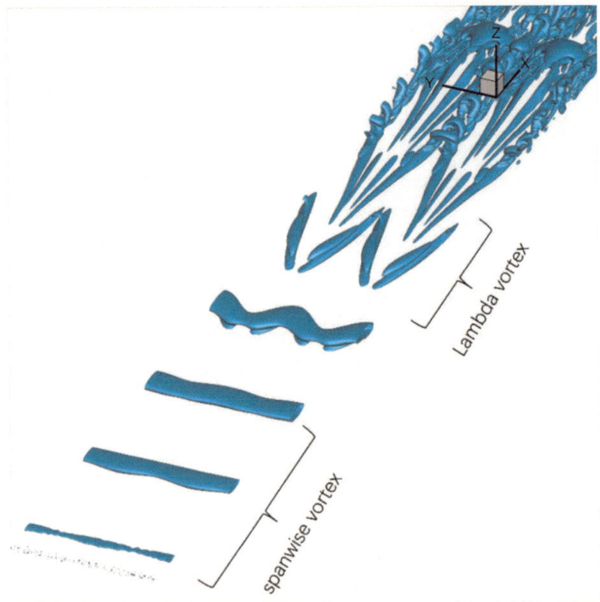


**Figure 11.**
*DNS result of the flat plate boundary layer transition.*

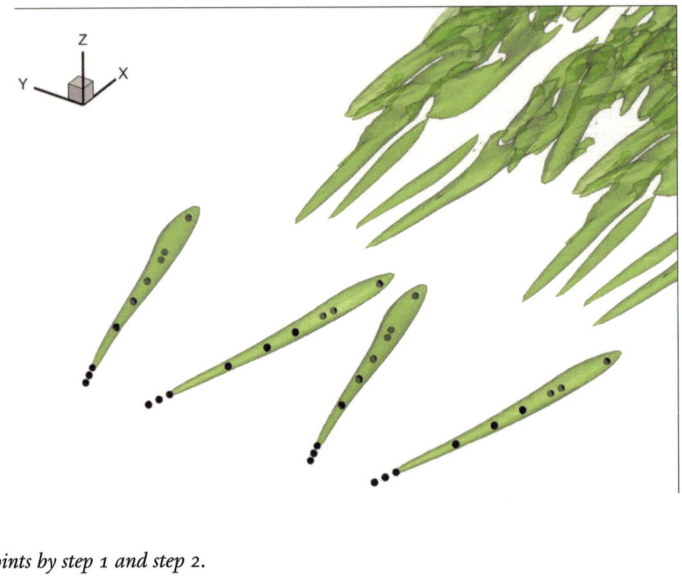


**Figure 12.**
*Selected seed points by step 1 and step 2.*

unique line. **Figures 21–23** show Liutex magnitudes at the vortex core positions by colors. They show the rotation strength of the vortex core line is not uniform and is varied at different locations. This is what people cannot get by using the iso-surface methods as iso-surface is a surface where the magnitudes are equal to the threshold.

**Figure 13.**
*(top view) Selected seed points by step 1 and step 2.*

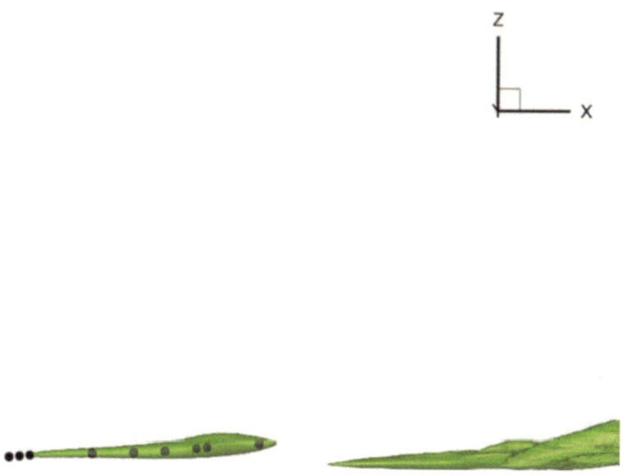


**Figure 14.**
*(side view) Selected seed points by step 1 and step 2.*

**Figure 15.**
*Selected seed points and corresponding forward Liutex gradient lines.*

**Figure 16.**
*(top view) Selected seed points and corresponding forward Liutex gradient lines.*

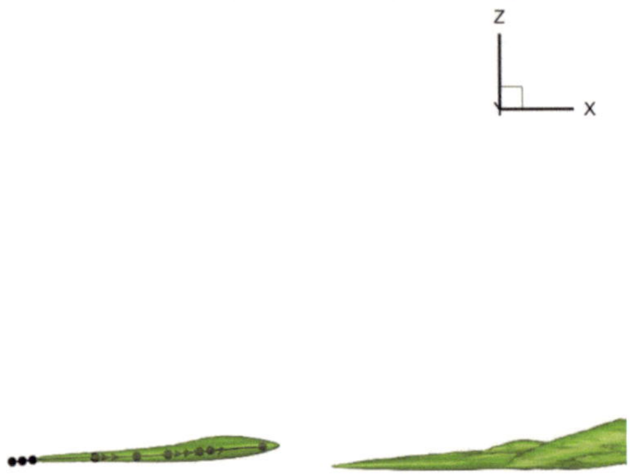


**Figure 17.**
*(side view) Selected seed points and corresponding forward Liutex gradient lines.*

**Figure 18.**
*Liutex lines generated by the same seed points.*

**Figure 19.**
*(top view) Liutex lines generated by the same seed points.*

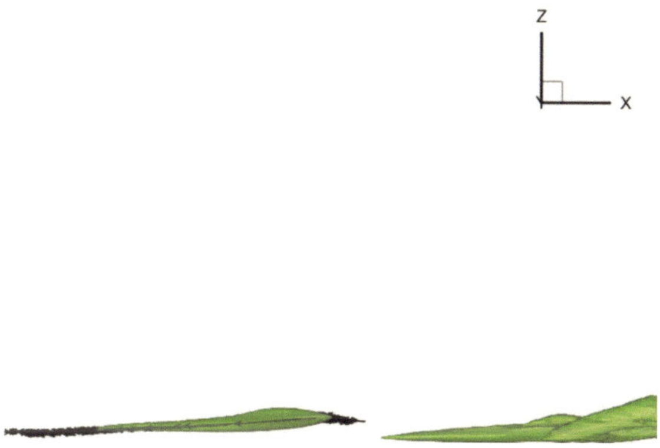


**Figure 20.**
*(side view) Liutex lines generated by the same seed points.*

## 7. Conclusion

This study reports a method to automatically draw Liutex core lines from a new perspective. The proposed method is based on the property of gradient lines that they are converging toward to the local maximum points. Instead of Liutex lines, Liutex gradient lines are used in the proposed method and give this method two major merits. Firstly, the uniqueness comes from the property of gradient lines. Secondly, the proposed method can tolerate some numerical errors as these errors can easily be eliminated during the process that gradient lines converge to the center. However, this is not a mature method, and some work still needs to be done in the future, for example, developing it into a universal method that works for different cases.

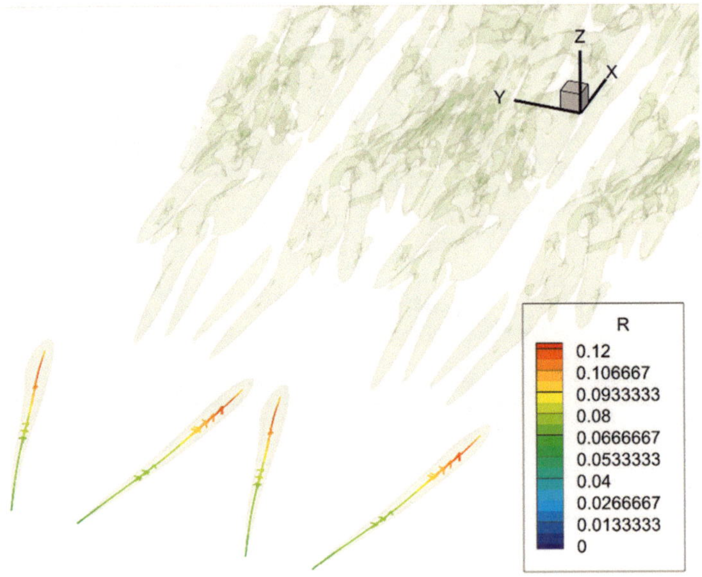


**Figure 21.**
*Liutex iso-surface and colorful forward Liutex gradient lines.*

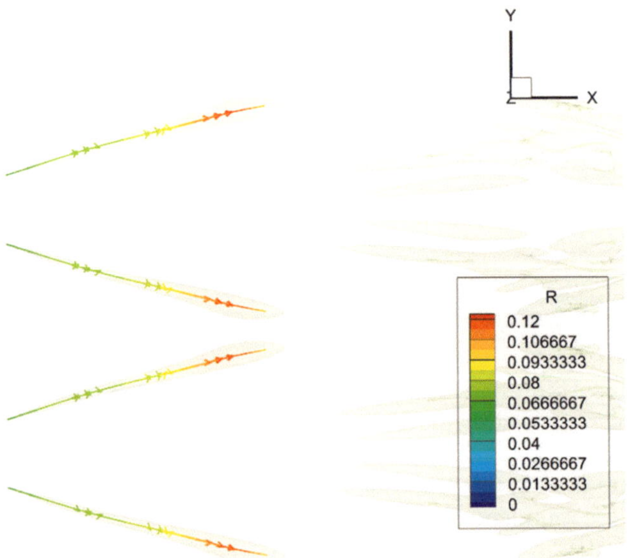


**Figure 22.**
*(top view) Liutex iso-surface and colorful forward Liutex gradient lines.*

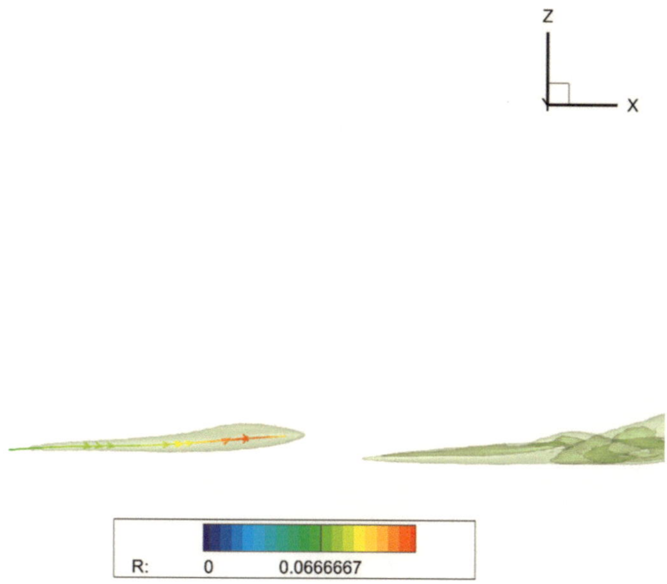


**Figure 23.**
*(side view) Liutex iso-surface and colorful forward Liutex gradient lines.*

## Acknowledgements

The project is currently supported by NSF under Grant #2300052. The authors are thankful for the support by the UTA Department of Mathematics, which houses the UTA Vortex and Turbulence Research Team. The authors are also grateful to Texas Advanced Computing Center (TACC) for providing computation time.

## Data availability

The data that supports the findings of this study are available from the corresponding author upon reasonable request.

## Author details

Yifei Yu and Chaoqun Liu*
Department of Mathematics, University of Texas at Arlington, Arlington, Texas, USA

*Address all correspondence to: cliu@uta.edu

## References

[1] Kaczorowski M, Chong K-L, Xia K-Q. Turbulent flow in the bulk of Rayleigh–Bénard convection: Aspect-ratio dependence of the small-scale properties. Journal of Fluid Mechanics. 2014;**747**:73-102

[2] Xi H-D, Zhang Y-B, Hao J-T, Xia K-Q. Higher-order flow modes in turbulent Rayleigh–Bénard convection. Journal of Fluid Mechanics. 2016;**805**:31-51

[3] Zhou Y. Rayleigh–Taylor and Richtmyer-Meshkov instability induced flow, turbulence, and mixing. I. Physics Reports. 2017;**720**:1-136

[4] Zhou Y. Rayleigh–Taylor and Richtmyer–Meshkov instability induced flow, turbulence, and mixing. II. Physics Reports. 2017;**723**:1-160

[5] Liu C, Gao Y-S, Dong X-R, Wang Y-Q, Liu J-M, Zhang Y-N, et al. Third generation of vortex identification methods: Omega and Liutex/Rortex based systems. Journal of Hydrodynamics. 2019;**31**:205-223. DOI: 10.1007/s42241-019-0022-4

[6] Wang Y, Yang Y, Yang G, Liu C. DNS study on vortex and vorticity in late boundary layer transition. Communications in Computational Physics. 2017;**22**:441-459. DOI: 10.4208/cicp.OA-2016-0183

[7] Hunt JCR, Wray AA, Moin P. Eddies, stream, and convergence zones in turbulent flows. Center for Turbulent Research Report CTR-S88. 1988:193-208

[8] Chong MS, Perry AE, Cantwell BJ. A general classification of three-dimensional flow fields. Physics of Fluids A: Fluid Dynamics. 1990;**2**:765-777. DOI: 10.1063/1.857730

[9] Zhou J, Adrian RJ, Balachandar S, Kendall TM. Mechanisms for generating coherent packets of hairpin vortices in channel flow. Journal of Fluid Mechanics. 1999;**387**:353-396. DOI: 10.1017/s002211209900467x

[10] Jeong J, Hussain F. On the identification of a vortex. Journal of Fluid Mechanics. 1995;**285**:69-94

[11] Gao Y, Liu C. Rortex and comparison with eigenvalue-based vortex identification criteria. Physics of Fluids. 2018;**30**:085107. DOI: 10.1063/1.5040112

[12] Cuissa JC, Steiner O. Innovative and automated method for vortex identification-I. Description of the SWIRL algorithm. Astronomy & Astrophysics. 2022;**668**:A118

[13] Shen C, Yang R, Wang M, He S, Qing S. Application of vortex identification methods in vertical slit fishways. Water. 2023;**15**:2053

[14] Xu W, Wang Y, Gao Y, Liu J, Dou H-S, Liu C. Observation on Liutex similarity in the dissipation subrange of turbulent boundary layer. Computers & Fluids. 2022;**246**:105613

[15] Xu W-Q, Wang Y-Q, Gao Y-S, Liu J-M, Dou H-S, Liu C. Liutex similarity in turbulent boundary layer. Journal of Hydrodynamics. 2019;**31**:1259-1262. DOI: 10.1007/s42241-019-0094-1

[16] Yan B, Wang Y, Liu C. Liutex-represented vortex spectrum in turbulence. Entropy. 2022;**25**:25

[17] Trieu XM, Liu J, Gao Y, Charkrit S, Liu C. Proper orthogonal decomposition analysis of coherent structure in a turbulent flow after a micro-vortex

generator. Applied Mathematical Modelling. 2022;**104**:140-162

[18] Dong X, Hao C, Liu C. Correlation between vorticity, Liutex and shear in boundary layer transition. Computers & Fluids. 2022;**238**:105371

[19] Dong X, Hao C, Dong Y, Liu C, Li Y. Investigation of vortex motion mechanism of synthetic jet in a cross flow. AIP Advances. 2022;**12**: 035045

[20] Liu C, Yu Y, Gao Y-S. Liutex based new fluid kinematics. Journal of Hydrodynamics. 2022;**34**:355-371

[21] Liu C, Liu Z. New governing equations for fluid dynamics. AIP Advances. 2021;**11**:115025. DOI: 10.1063/5.0074615

[22] Liu C. New ideas on governing equations of fluid dynamics. Journal of Hydrodynamics. 2021;**33**:861-866

[23] Yu Y, Shrestha P, Alvarez O, Nottage C, Liu C. Investigation of correlation between vorticity, Q, λci, λ2, Δ and Liutex. Computers & Fluids. 2021;**225**: 104977. DOI: 10.1016/j.compfluid.2021.104977

[24] Charkrit S, Shrestha P, Liu C. Liutex core line and POD analysis on hairpin vortex formation in natural flow transition. Journal of Hydrodynamics. 2020;**32**:1109-1121

[25] Yu Y, Shrestha P, Alvarez O, Nottage C, Liu C. Correlation analysis among vorticity, Q method and Liutex. Journal of Hydrodynamics. 2020;**32**: 1207-1211. DOI: 10.1007/s42241-020-0069-2

[26] Dong X-R, Cai X-S, Dong Y, Liu C. POD analysis on vortical structures in MVG wake by Liutex core line identification. Journal of Hydrodynamics. 2020;**32**:497-509

[27] Wang Y-Q, Gao Y-S, Xu H, Dong X-R, Liu J-M, Xu W-Q, et al. Liutex theoretical system and six core elements of vortex identification. Journal of Hydrodynamics. 2020;**32**:197-211. DOI: 10.1007/s42241-020-0018-0

[28] Yu Y, Shrestha P, Nottage C, Liu C. Principal coordinates and principal velocity gradient tensor decomposition. Journal of Hydrodynamics. 2020;**32**: 441-453. DOI: 10.1007/s42241-020-0035-z

[29] Tran CT, Long X-P, Ji B, Liu C. Prediction of the precessing vortex core in the Francis-99 draft tube under off-design conditions by using Liutex/Rortex method. Journal of Hydrodynamics. 2020;**32**:623-628

[30] Wang Y, Gao Y, Liu C. Liutex and Third Generation of Vortex Identification:Workshop from Aerospace and Aeronautics World Forum 2021. Singapore: Springer Nature; 2023

[31] Liu C, Wang Y. Liutex and Third Generation of Vortex Definition and Identification: An Invited Workshop from Chaos 2020. Cham: Springer Cham; 2021

[32] Liu C, Xu H, Cai X, Gao Y. Liutex and Its Applications in Turbulence Research. London: Academic Press; 2020

[33] Liu C, Gao Y. Liutex-Based and Other Mathematical, Computational and Experimental Methods for Turbulence Structure. Singapore: Bentham Science Publishers; 2020

[34] Gao Y-S, Liu J-M, Yu Y-F, Liu C. A Liutex based definition and identification of vortex core center lines.

Journal of Hydrodynamics. 2019;**31**: 445-454. DOI: 10.1007/s42241-019-0048-7

[35] Xu H, Cai X-S, Liu C. Liutex (vortex) core definition and automatic identification for turbulence vortex structures. Journal of Hydrodynamics. 2019;**31**:857-863. DOI: 10.1007/s42241-019-0066-5

[36] Li H, Liu Y, Wang D, Xu H. Liutex (vortex) core and tube identification and automatic generation algorithms. Computers & Fluids. 2023;**250**:105731. DOI: 10.1016/j.compfluid.2022.105731

[37] Yan Y, Liu C. Shear layer stability analysis in later boundary layer transition and MVG controlled flow. 51st AIAA Aerospace Sciences Meeting including the New Horizons Forum and Aerospace Exposition. 2013

[38] Bake S, Meyer D, Rist U. Turbulence mechanism in Klebanoff transition: A quantitative comparison of experiment and direct numerical simulation. Journal of Fluid Mechanics. 2002;**459**:217-243